Imperiled Reef

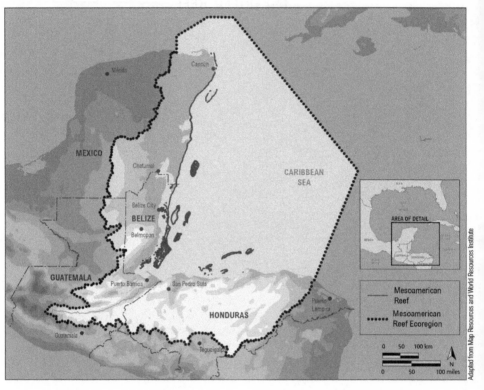

Mesoamerican Barrier Reef Ecoregion, courtesy of the Healthy Reefs Initiative.

Healthy Reefs
for healthy people

IMPERILED REEF

The Fascinating, Fragile Life of a Caribbean Wonder

SANDY SHEEHY

UNIVERSITY OF FLORIDA PRESS
Gainesville

26 25 24 23 22 21   6 5 4 3 2 1

ISBN 978-1-68340-249-7
Library of Congress Control Number: 2021930613

University of Florida Press
2046 NE Waldo Road
Suite 2100
Gainesville, FL 32609
http://upress.ufl.edu

UF PRESS

UNIVERSITY
OF FLORIDA

In memory of my father, Robert Granville,
who nurtured my curiosity
about the natural world

CONTENTS

PREFACE

I first became interested in the undersea world when I was four or five years old. I remember sitting in the waiting room of my pediatric dentist's office, fascinated by the colorful little fish in the tropical aquarium complete with miniature orange castle. When I was a bit older, my favorite television show was *Sea Hunt*, the 1950s series of thrillers in which Lloyd Bridges, best known nowadays as the father of actors Beau and Jeff, played former U.S. Navy frogman Mike Nelson, who fought bad guys beneath the waves. A decade later, I became a fan of the TV documentary series *The Undersea World of Jacques Cousteau*, which pioneered a new level of sophistication in underwater cinematography. As an adult, I noticed that many of the fantastic creatures on the planet Pandora in James Cameron's 2009 film *Avatar* were inspired by reef species here on Earth.

My father taught me how to swim when I was three, so proceeding to snorkeling was a natural for me; even a lost hairband on the bottom of a swimming pool was reason enough to look beneath the surface. After a couple of trips to Florida and the Caribbean, I got my NAUI scuba certification, so that I could swim among the creatures on the coral, rather than merely watching them from above.

Let me explain right up front that I am not an underwater photographer. All photos that appear in this book are black and white, contributed by the individuals credited in the captions. Libraries, online outlets, dive shops and the remaining bricks-and-mortar bookstores carry stunning books of full-color underwater photographs, from coffee-table tomes to pocket guides with their pages thoughtfully waterproofed so that they can be carried aboard a dive boat without becoming soggy. Some, such as Underwater Editions' *Cozumel: Dive Guide and Log Book*, spiral bound for handy reference on board, focus on one particular part of the reef. *Reef*

Creature in-a-Pocket: Florida, Caribbean, Bahamas by Paul Humann and Ned DeLoach is both wider-ranging and waterproof. A classic of the large format genre is *Coral Gardens*, the lushly photographed 1978 volume by German actress and director Leni Riefenstahl, better known for the Nazi propaganda film *Triumph of the Will* and for her documentary of the 1936 Summer Olympics. A more recent book, *Citizens of the Sea: Wondrous Creatures from the Census of Marine Life*, published in 2010 by the National Geographic Society, is rich in remarkable photos, as well as surprising details, of life beneath the world's waves. Its author, Nancy Knowlton, founded the Center for Marine Biodiversity and Conservation at the University of California San Diego's Scripps Institution of Oceanography and is Sant Chair Emerita of Marine Science for the Smithsonian Institution's Museum of Natural History.

I own a number of handsome, informative underwater photo books and nature guides that I list in the bibliography. I am particularly fond of Hofstra University biologist Eugene Kaplan's *Coral Reefs: A Guide to the Common Invertebrates and Fishes of Bermuda, the Bahamas, Southern Florida, the West Indies, and the Coast of Central and South America*, part of the Roger Tory Peterson series of field guides sponsored by the National Audubon Society and the National Wildlife Federation. My copy happens to be leather-bound, complete with gilt-edged pages and a pretty orange ribbon attached as a bookmark, so it doesn't accompany me on dive boats, but the arrangement of the material is sensible, the text is thorough and accessible, and the photos (supplemented by Susan Kaplan's illustrations) have proved more useful for making sense of what I see in the briny deep than those in sturdier but briefer guides.

As a journalist, rather than a marine biologist or oceanographer, I am deeply indebted to the knowledge that specialists in these and related fields have shared so generously in books, in theses, in journal articles and conference proceedings, and through interviews and correspondence. I am particularly grateful to Zach Foltz, station manager and dive officer for the Smithsonian Institution's Caribbean Coral Reef Ecosystems Program, who arranged for me to spend a week at the Smithsonian's Carrie Bow Cay Field Station on the Belize section of the Mesoamerican Barrier Reef. There I was privileged to observe Department of the Interior biological oceanographer Karen Koltes and her colleagues as they monitored the health of the coral colonies and of the seagrass beds that lie between the reef and the mainland, nourishing and protecting hundreds of species of juvenile fishes and crustaceans. Witnessing the meticulous day-to-day tasks of marine field work gave me immense admiration for those who conduct it.

I am also especially thankful to environmental engineers Ian Drysdale, Honduras in-country coordinator for the Healthy Reefs Initiative, and Jennifer Myton, associate program director for the Coral Reef Alliance (CORAL), who shared their perspectives on the effective community-based projects to improve and protect the health of the Mesoamerican Barrier Reef.

In reviewing the manuscript in draft, marine biologist, invertebrate zoologist and conservation ecologist Richard Brusca, coral reef ecologist Stephen Gittings and marine biologist Donald Behringer provided valuable suggestions and caught what would have been embarrassing mistakes.

As this book goes to press, the COVID-19 pandemic continues to impact the countries along the Mesoamerican Barrier Reef, both by infecting residents and by discouraging tourism. The effects described in this book reflect early 2021 figures.

Any errors or omissions in the following chapters are solely my own.

The author dressed for diving, photo by Charles McClelland.

INTRODUCTION

The best way to understand and appreciate the natural world that surrounds us may be to explore beneath its surface—not by digging, which destroys the very systems we seek to examine, but by looking underwater, which we can do with little or no disruption of the processes we observe. Birth and death, growth and decline, the interaction of plants, animals and minerals, of predators and prey—in short, everything that happens in the familiar environment aboveground also happens below the waves. Because water is transparent, these phenomena should be easier to observe there; yet the environment is so profoundly alien to human beings that venturing down even 40 or 50 feet is the closest almost any of us will come to visiting another planet.

A person needn't be a scuba diver or a snorkeler to explore the fascinating complexity of life beneath the surface. You don't even have to be able to swim. I prefer to visit this engagingly exotic world in person, listening to the gentle bubbles of my exhalations as I glide lazily along, enveloped in warm liquid. (As a diver, I'm something of a wuss. I never was one for cold-water diving and have a decided preference for the tropics.) But I have become increasingly amazed at the quality and detail captured by the world's great underwater cinematographers and even by enthusiastic graduate students. The latter's camera work may waver, but their obvious excitement and their exclamations of "Awesome grouper!" and the like make up for the home-movie visual quality.

Life on earth began in the sea four billion years ago.[1] Oceans now cover 127,970 square miles of the earth's surface.[2] Combined they contain 300 million trillion gallons. Each drop of seawater holds between 2,500 and 300,000 single-celled organisms—bacteria, viruses, protists, and protozoans—taken together, a hundred million times more than the estimated number of stars in the universe.[3]

As oceanographer Sylvia A. Earle put it poetically in her 2009 book *The World Is Blue:* "Seen from a space shuttle, Earth's continents seem to be islands floating in a shimmering indigo embrace ..."[4]

That "shimmering indigo" embraces millions of species. According to the National Oceanic and Atmospheric Administration, only 9 percent of the world's marine species have been catalogued. One reason may be that 80 percent of the ocean has yet to be explored.[5]

The roster of species is growing constantly, thanks to the development of genetic sequencing in the latter part of the twentieth century. Earle described the work of marine biologist Richard Pyle, who explored the ocean 330 to 660 feet down, finding 12 to 13 new species *per hour*.[6]

For all forms of life, scientific taxonomy proceeds from broadest to most specific: domain, kingdom, phylum, class, order, family, genus, species. In this book, I use common names for the plants and animals I discuss but also provide, at least once, their scientific names, in the form of genus and species—for example: hawksbill sea turtle (*Eretmochelys imbricate*); or, when the common name refers to members of a family, subfamily or genus, that more inclusive name—for example: blennies (family: Blenniidae).

Some scientists estimate that 25 percent or more of marine species depend for at least part of their life cycles on coral reefs,[7] although such reefs occupy only 0.1 percent of the oceans' total surface area and a thin layer at that; the average depth of the world's seas is 2.5 miles, and the deepest waters reach down seven miles.[8] Situated between a foot or two below the surface and about 300 feet down, coral reefs rely on the filtered sunlight that algae and other aquatic plants require for photosynthesis, which in turn fuels the ever-looping chain of life for corals, sponges, crustaceans, eels, marine mammals, and fishes large, small and tiny.

Dotting the earth's shallow seas, coral reefs extend generally to 30 degrees latitude on either side of the equator. In some areas where waters farther north or south are unusually warm—such as Bermuda, located at 32.3 degrees north but bathed by the Gulf Stream—patch reefs of hard corals survive.[9]

Coral reefs come in three major varieties: fringe, which lie close to the shore or around islands; atoll, which started as fringe reefs but became more or less circular freestanding structures when their central islands subsided; and barrier, which form along continental shelves and around groups of islands. (When sea levels drop, islets can also emerge from now-exposed barrier reefs.[10]) For the purposes of this book, I decided to focus on one particular reef, the Mesoamerican Barrier Reef, also known as the Great Mayan Barrier Reef, in the Western Caribbean. I chose it for three reasons. First, although coral reefs occupy only 0.1 percent of the surface area of the world's oceans, as mentioned previously, that was still too much territory for me to cover in my remaining lifetime. Second, I already had a passing acquaintance with this particular stretch of coral; I'd been diving parts of it for decades. Third, all the beauty and drama of the coral ecosystem, and all the forces and follies that imperil coral reefs everywhere, were clearly evident there.[11]

Resting on a foundation of limestone and hard corals, at 625 miles long the Mesoamerican Barrier Reef measures less than half the length of Australia's Great Barrier Reef; yet in natural beauty and ecological interest, it rivals its more famous cousin. Stretching from Isla Contoy, just north of Cozumel, Mexico's largest Caribbean resort island, to the Bay Islands of Honduras—Roatán, Útila, and Guanaja, plus five smaller islands and 53 cays—the Mesoamerican Barrier Reef is the second-longest continental coral structure on earth.[12] As its official name reflects, this underwater paradise is a multinational asset, skirting the coasts of Mexico, Belize, Guatemala, and Honduras, lying between a few hundred yards and 22 miles offshore.

Four atolls, rings of islets that started as fringe reefs surrounding a volcano but now rise barely above the surface, belong to the barrier reef system: Chinchorro Banks, Turneffe Atoll, Lighthouse Reef, and Glover's Reef.[13] (Encompassing five islets in the southern crescent of the Gulf of Mexico, 60 miles north of the Mexican port of Progresso, Arrecife Alacránes, or Scorpion Reef, sits more than 200 miles to the west of the northern tip of the Mesoamerican Barrier Reef. Some scientists argue that Scorpion Reef is not technically an atoll because its origin is not volcanic.[14])

Much has been made about coral reefs in recent years, especially about the perils they face due to global warming. The hard corals that form the foundation for the reef ecosystem are very particular about salinity and temperature. Most species do best in water between 73°F and 84°F with a salt concentration between 32 and 42 parts per thousand.[15] If the warmth of the water they inhabit falls below 64°F for more than a few hours, they die. Although some species can tolerate temperatures as high as 104°F for brief periods, most expel their Symbiodiniaceae, the family of algae with which healthy corals live symbiotically (and which lend them their vivid colors), if the water temperature rises above 90°F.[16] Called "bleaching," this visually striking symptom indicates a patch of coral under extreme stress and vulnerable to lethal damage from storms and disease.

This book does address bleaching and the other threats to the Mesoamerican Barrier Reef and its cousins worldwide. Changing acidity of the ocean, nutrient-laden runoff from fertilizer, sediment kicked up by storms, debris and sewage from coastal development, introduction of invasive species, overfishing, inept divers, and plain old trash, such as plastic soda bottles, all these and more endanger this fragile ecosystem, as well as the other 99.9 percent of the earth's oceans. Most of these threats result from human activity. Some can be halted or at least curbed relatively easily, given broader awareness, targeted economic incentives and international cooperation. The concluding chapters of this book describe specific examples of successful programs and specific ways in which we

can act, collectively and individually, to limit and even reverse the damage.

But people don't protect what they don't value. The purpose of this book is to raise understanding and appreciation of one particular, fascinating and fragile ecosystem and thus, by extension, of coral reefs throughout the tropical oceans and of the rest of the earth's environment everywhere.

Healthy coral ecosystem, photo by Ben Phillips.

I

THE ESSENTIAL REEF

THE ESSENTIAL KEEP

1

HOW TO BUILD A BARRIER REEF

Forget the familiar portrait of Charles Darwin as an aged legend, with his fringe of white hair, his bushy eyebrows and his full beard reaching several inches below the top V of his waistcoat; and imagine instead a man in his late 20s, his body fit, his eyes alight with curiosity.

Darwin was a competent draftsman (a skill common to scientists, as well as engineers and military officers, of his day) and a meticulous observer. Examining Keeling Atoll, now better known as the Cocos Islands, in the Indian Ocean due south of Sumatra, he noted that the sections of coral subjected to the constant sunlight at the surface died. Deciding that this phenomenon deserved a closer look than that afforded from the deck of the *Beagle*, he pole-vaulted onto this jagged reef top, risking painful scrapes to stand within a few yards of the crashing breakers. His account reads like a nineteenth-century adventure story:

> [I]t is possible only under the most favorable circumstances, afforded by an unusually low tide and smooth water, to reach the outer margin [of the reef], where the coral is alive. I succeeded only twice in gaining this part, and found it almost entirely composed of a living Porites [stony finger coral], which forms great irregularly rounded masses . . . from four to eight feet broad, and little less in thickness. These mounds are separated from each other by narrow crooked channels, about six feet deep; most of which intersect the line of reef at right angles. *On the furthest reef, which I was able to reach by the aid of a leaping-pole, and over which the sea broke with some violence* [italics mine], although the day was quite calm and the tide low, the polypifers in the uppermost cells were all dead, but between three and four inches lower down on its side, they were living, and formed a projecting border around the upper and dead surface.[1]

Then, as the breakers crashed at his feet, the young Darwin took soundings off the far edge of the reef using a bell-shaped lead sinker about four inches wide at its concave bottom, to which he attached a wad of tallow to retrieve impressions of the coral. Beginning near the surface, he dropped the lead successively deeper, noting that the tallow almost always came up smooth above 72 feet.[2] Below that, the bottom consisted of sand or water-worn rock.

Noticing that coral larvae only attached themselves to rocks or other hard structures in the relative shallows, Darwin concluded that after a volcano burst through the surface, forming an island, the larvae, spawned by corals far away and hitching a ride on ocean currents, would settle into the sunlit water next to the land. There they would build what he called a fringe reef. Eventually, the island would erode or subside, creating a placid lagoon between itself and what would now be a barrier reef. In time, the island would disappear beneath the surface, and the roughly circular reef that had surrounded it would become an atoll.

Based on these observations, in 1842 Darwin published what was for its time the definitive work on coral reefs. Appearing 17 years before *On the Origin of Species*, *The Structure and Distribution of Coral Reefs* was the iconic naturalist's first publication. The scientific community of the day received it enthusiastically, and the major findings it brought forth remain unchallenged to this day,[3] which is especially remarkable given that Darwin conducted all his observations from the surface. He lacked scuba or even snorkeling gear. Pearl- and sponge-divers had been plunging into the depths since ancient Greece, divers unaided by anything but a heavy rock to help them descend had salvaged valuables from wrecks since seafaring began, and the diving bell had been in use at least since 1531, when Italian inventor Guglielmo de Lorena utilized one to recover treasure from submerged Roman ships. A rudimentary version, perhaps an inverted cauldron, was probably used well before that, perhaps as early as the fourth century BCE. Legends and epic poems describe Alexander the Great employing a glass barrel to allow him to breathe underwater during the Siege of Tyre.[4] In 1775 Scottish candy-maker Charles Spalding improved on the concept with a sort of submersible elevator.[5] However, these methods didn't allow the luxury of scientific exploration. It wasn't until 1943, more than a century after Darwin's breakthrough book, that Jacques-Yves Cousteau and Émile Gagnan patented a functional self-contained underwater breathing apparatus, giving the world the acronym "SCUBA."[6] Able to maneuver untethered beneath the surface, researchers henceforth could study coral reefs up close and at varying depths.

Darwin made his observations in the Pacific and Indian Oceans, during his voyage on the HMS *Beagle* from 1831 to 1836.[7] The closest he got to Central America was Brazil on the Atlantic side and the Galapagos Islands, off Ecuador, on the Pacific, so he missed the Mesoamerican Barrier Reef.[8] Darwin freely admitted that his descriptions of the coral reefs of the West Indies were drawn from the maps and charts of other maritime explorers.[9] Nonetheless, he did examine the Great Barrier Reef off Australia (then known as New Holland) and the smaller one in New Caledonia, and his basic conclusions apply to coral barrier reefs in tropical waters everywhere.

I knew of Darwin as the author of *On the Origin of Species,* the naturalist aboard the HMS *Beagle,* and the father of the theory of evolution. I had read *To the Edge of the World,* Harry Thompson's riveting account of Darwin's voyage around South America. But I had never heard about his pioneering work on coral reefs until Thomas M. Iliffe, Professor of Marine Biology at Texas A&M University at Galveston, told me about it in 2016. Iliffe was a middle-aged man with a receding hairline and a neatly trimmed beard and moustache. We were sitting in his modest office, the two visitors' chairs wedged between his desk and the back wall. The single window looked out at the narrow dock where several of the university's research boats lay in wait for the students, called "Sea Aggies," who would return the following week for the start of the spring semester.

According to Iliffe, what Charles Darwin had concluded in the 1840s was still the best model of coral reef formation around islands and held, with one important difference, for coastal barrier reefs.[10] Describing the probable genesis of the Mesoamerican Barrier Reef, Iliffe explained that unlike barrier reefs that ring islands in the Pacific and Indian Oceans, but like its larger cousin off Australia, it didn't form around a volcano.[11] It probably began as a coastal fringing reef during the last interglacial period (about 127,000 to 116,000 years ago), when the surface of the Caribbean Sea was 20 to 30 feet higher than it is today.[12] Iliffe described finding an underwater cliff indicating that the shoreline of the Yucatán was 413 feet below its present level during the Last Ice Age about 20,000 years ago.[13]

Human beings inhabited caves in that limestone and the passages that connected the inland sinkholes called cenotes. Well-developed stalactites hang from the ceilings and stalagmites rise from the floors of these labyrinths, clear evidence that they once were dry.[14] According to findings by Mexican underwater archeologist Arturo González González and his German colleague Wolfgang Stinnesbeck, the first wave of settlers to make these chambers their homes arrived in the Yucatán as early as 13,000 years ago, probably crossing from Asia via the Bering Strait land bridge, then over the course of thousands of years making their way south and east.[15]

Beginning in the first part of the present century, underwater archeologists have discovered in now-flooded caves the remains of species, such as the American horse and a New World elephant, wiped out during the massive extinction at the end of the Last Ice Age. A bone of one, a camelid related to the present-day llama, was charred and left next to a pile of ashes, indicating that the animal had been cooked and eaten.[16]

And underwater archeologists have found human remains, as well. The DNA of the oldest proved to be closer to samples retrieved from burial sites hundreds of miles away in the state of Puebla and even to those from China than to the DNA of present-day Mayas—or to that of pre-Columbian Maya remains recovered from cenotes.[17] Before these more recent discoveries, which included a skeleton 11,600 years old discovered by González and Stinnesbeck,[18] scientists had thought that humans first arrived at the peninsula between 4,000 and 2,000 years ago.[19] The jungle had degraded any evidence older than the start of the Common Era, but the fresh water and fine silt of the sinkholes and caves had preserved far more ancient artifacts.[20]

As to those Mayan remains, underwater archeologists have also turned previous wisdom on its head. Initially convinced that the Mayan monumental structures in the Yucatán jungle were the remains of the legendary lost civilization of Atlantis,[21] early twentieth-century Massachusetts Mayanist Edward Herbert Thompson bought the plantation that included Chichén Itzá, a 1,500-year-old site that surrounded a sinkhole dubbed the Sacred Cenote. Between 1904 and 1909, Thompson dredged the cenote, bringing up ceramics, gold and jade jewelry, and human remains bearing evidence of wounds that indicated that they had been sacrificed. Because sixteenth-century Spanish missionaries had documented such rituals, and banned them, this was a logical interpretation; and it was at least partially correct.[22] Neither Thompson's crude methods nor his habit of shipping relics of Mexico's past to Harvard's Peabody Museum would pass muster today. But although he lacked the underwater technology to examine human remains and artifacts in situ, he went about his research seriously, seeking to understand the civilization that had once inhabited the Yucatán and had disappeared so mysteriously. (Thompson eventually concluded that Chichén Itzá was not Atlantis.[23])

Nine decades later, recreational cave divers exploring the cenotes began reporting encountering human remains and artifacts. Some of these dated from the Spanish colonial period, but the oldest were prehistoric.[24] In 1999 Mexico's Instituto Nacional de Antropología e Historia began efforts to protect and recover them. One of the cenotes in the state of Quintana Roo contained the remains of 118 individuals—so many that it was called Las Calaveras (The Skeletons).[25] In the 25 prehistoric sites and the dozens of Mayan sites discovered in the first six years of the program, many of the skeletons were 80 to 90 percent intact, indicating that the individuals had been laid out in funerary fashion, not tossed into the water to drown, their remains carried along underwater passages and eventually broken apart.[26] Some of the skulls displayed the distinctive sloping forehead resulting from the pre-Columbian Maya practice of binding young children's heads, suggesting that the Mayas had utilized the caves as burial sites at a time when they were dry.[27] Given that the Mayas considered cenotes and caves as occupying a boundary between the inhabited surface and the underworld, it would have made sense for them to use them as burial, as well as sacrificial, sites.

As Paleolithic people settled the slab of Yucatán limestone (which, after all, included layers of ancient seabeds), just offshore a fringe reef was forming. The first corals to colonize what is now the Mesoamerican Barrier Reef attached themselves to rock now 7,000 years old.[28] Because the seas were lower, that rock was close to the surface. As the seas rose, the original coral colonists died; new ones built on their exoskeletons—very, very gradually. In this part of the Caribbean, the rate of reef growth upward is only about one to six yards per thousand years.[29]

The Mesoamerican Barrier Reef began as a fringe reef, its crest close enough to the surface to provide the sunlight that the corals' symbiotic algae required to survive. As the sea rose and the mainland subsided, the hard corals continued to grow upward toward the light. The lagoon between the barrier reef and the coast developed, along with the peculiar geology of the Yucatán. The northern two-thirds of the Yucatán Peninsula is very old karst—limestone that began as coral reef. Lying near the coast, the youngest limestone is 120,000 years old.[30] Except for wetlands at the northeast end, not until well south of a line running from Campeche on the peninsula's west coast to Punta Soliman on the east do streams begin to emerge on the jungle floor; and those that do surface run south, to inland lakes and to the Bay of Chetumal on the Mexico-Belize border, not east to the sea. Throughout most of the peninsula, rainwater percolates vertically through the limestone, eventually finding an underground river that takes it to the Caribbean.[31]

Off the Belize sector of the Mesoamerican Barrier Reef, five submarine ridges parallel Belize's major rivers. The best-developed ridge extends north, forming the foundation for two atolls, Glover's Reef and Lighthouse Reef.[32] The shelf in Belize extends to the east, sloping gradually to 200 yards in depth, then dropping off at a 40-degree angle into the abyss.[33] The depth on the east side reaches 9,000 feet.[34]

"Why the reef is where it is is a good question," Gilbert Rowe, Texas A&M University at Galveston regents professor of marine biology, explained as we sat in his office down the hall from Iliffe's. "The ocean is full of larvae, whether it's a fish or a small worm. Wherever they find habitat, they stay."[35]

Oceanographers have a good idea why the Mesoamerican Barrier Reef ends so abruptly at Isla Contoy: the ocean currents that roil past the northern tip of the Yucatán Peninsula.

"Isla Contoy is very interesting," observed Nuno Simoes, a marine biologist at the Yucatán campus of the Universidad Nacional Autónoma de México. "In the south part, you find corals. In the north part, you have very cold, murky waters and almost no coral. You go just 500 meters or a kilometer, and it's a completely different environment."[36]

The reason? A huge upwelling of cold water in the north.[37]

Why so many larvae chose the particular stretch of marine outcrop running south from Isla Contoy to the Bay Islands of Honduras may be a mystery, but the conditions that made growth possible were present at the time—and may not be in the near future. The Mesoamerican Barrier Reef was able to develop where it did because of water temperature, and thus latitude.[38]

"If there's water temperature below 20 degrees centigrade (around 70 degrees Fahrenheit) for prolonged periods, you aren't going to find reef-building corals," Rowe clarified. "You don't find coral reefs farther from the equator than the 20 centigrade isotherm."[39]

Noting that temperatures about 30°C, such as those in the Persian Gulf, may kill corals, he continued: "The main reason that the Caribbean doesn't get hotter than 28 centigrade is *evaporation*. There's a lot of circulation in the Caribbean, which is an extension of the North Atlantic, and a lot of water turnover; so it doesn't get too salty, despite evaporation."[40]

Coral reefs rank as the world's most diverse ecosystems. Scientists estimate that they are home to as many as nine million species,[41] and thanks to DNA sequencing that number is increasing constantly. As Darwin noted, "The almost universal law of 'consume and be consumed'" applies to all tropical coral reefs.[42] But so do the rules of symbiosis and natural cooperation.

Each class of creatures and plants native to the reef helps maintain it. Colloquially known as Zooxanthellae, Symbiodiniaceae, the diverse family of algae that live symbiotically with hard corals and give them their color, utilize photosynthesis to provide energy-giving sugars to their hosts. Moray eels share hunting duties with Nassau groupers, flushing prey from niches in the reef, enabling their partners to grab it swimming. Using their sucker-equipped mouths, remoras clean parasites from sharks.

On the Mesoamerican Barrier Reef, as in seas around the world, each species depends on others. Microscopic organisms, marine plants, and algae use chlorophyll, carbon dioxide, and water to convert sunlight to oxygen, which enriches not just the seas but also the air we terrestrial beings breathe, and simple sugars, which pass to the animals that consume the ingredients of the diverse floating soup collectively known as plankton. Most of these plankton eaters are tiny, often larvae and juveniles of fish and invertebrates; but some, such as whale sharks, rank among the earth's giants. By eating the larvae and juveniles, as well as members of small species, or by grazing on aquatic plants or coral polyps, adults concentrate the energy and transfer it to whatever eats *them*.[43]

Thus, when one kind of creature goes missing from the reef, whether from disease, overfishing or a change in the water chemistry, the entire ecosystem can buckle.

Animals high on the food chain are especially vulnerable to increases in salinity. Jellyfish, for example, may easily accommodate a rise in salinity; but sea turtles, which eat jellyfish, concentrate their prey's salt in their own bodies.[44] Although turtles possess the means to expel the excess, even through what look like tears, these organs can become overwhelmed if lower-salinity water doesn't enter the system. Think of the landlocked Great Salt Lake, lifeless except for brine shrimp and certain saline-loving species of algae.

In addition to salinity another limiting nutrient is nitrate. You don't find coral reefs in parts of the ocean where nitrate "fertilizes" lots of tiny plants and animals. If the level of nitrates is low, that's good for corals, because they need clear water. If there's too much nitrate, there's a plankton bloom and the corals don't get enough light.[45]

As we sat in his office at Texas A&M University at Galveston, Tom Iliffe explained that the ecology of the Mesoamerican Barrier Reef has six components: mainland, mangrove swamps, beach, lagoon, reef crest, and deep reef, or fore reef.[46] The mainland consists primarily of dense jungle, where vines climb tropical hardwoods; iguanas, snakes, and crocodiles slither; birds with brilliant feathers flit; anteaters and armadillos root; monkeys chatter; and small mammals with names like coati and grison try to avoid becoming late-night snacks for jaguars. At some points, towns, pastures, cultivated fields, and orange groves tear through this vibrating green mat; but left to nature it segues into brackish ponds and marshes, where tight clumps of mangroves filter the nutrient-rich silt and add the leaves and bark they shed. The resultant mud helps feed the marine plants and animals near shore, thus exporting material from the mainland to the reef.[47]

Short, broadleaf trees with prominent roots, mangroves are remarkable for their tolerance of the salty, low-oxygen water in which they grow. Although mangroves cover only 0.5 percent of the earth's tropical coasts, they shoulder between 10 and 15 percent of the carbon storage;[48] by some accounts they are six times as effective as rain forests at absorbing carbon.[49] They also prevent erosion and serve as a nursery ground for species that will spend their adult lives offshore.[50]

Beaches are another story. Except in coves and bays, only a narrow fringe of sand, if any, lies between the mangrove roots and the water. Some of that sand has passed through the intestines of parrotfish (family: Scaridae, relatives of wrasses) and other creatures that graze on corals. Parrotfish digest the coral polyps, then excrete the indigestible bits of exoskeleton in finely ground, beautifully white form.[51] Some of that dazzling bit of beach on which tourists bask may well have begun as the feces of thousands upon thousands of parrotfish.

Below the surface, the beach becomes sand flats, where seagrass shelters juvenile fish, crustaceans, and turtles.[52] Little bivalves like the winged pearl oyster cling to the blades.[53] Otherworldly looking creatures like sea slugs and sea hares crawl along the sand. Lacking shells, they sport spots and protuberances like psychedelic locomotives. The most bizarre, and often the most beautiful, of these mollusks are the nudibranchs. Some camouflage themselves by blending into the underwater landscape, while others (the ones we notice) sport vivid colors and feathery flourishes to warn hungry fish that the merest bite might taste nasty and may even pack a toxic punch. Nudibranchs create both their striking colors and their foul flavors by ingesting sponges, sea anemones, jellyfish, and other marine creatures that have developed chemical defenses against predators, then concentrating those substances in their own bodies. These nasties don't bother the nudibranchs.[54]

As the water deepens into a lagoon, a relatively placid environment, adult conch dot the sandy bottom, stingrays drift along, and garden eels wave in the current, mimicking aquatic plants until approached by a curious diver, when these startling creatures duck back into the sand. Brown file clams burrow into the bottom.[55] By circulating water through cuts on the seaward side, tides keep the lagoons from becoming stagnant.

Coastal lagoons along the Mesoamerican Barrier Reef range in width from 8 miles along the northern Yucatán to 30 miles off the coast of Belize.[56]

To have a lagoon, you have to have a barrier protecting it from the open ocean. Along the East Coast of the United States, this protection is provided by islands of sand tossed up by the surf. In the Caribbean, it comes from barrier reefs.

When water levels drop, exposing colonies of reef-building corals, the polyps die and the exposed sections of reef form islands; but these are different from the sand islands along the Atlantic. The islands of the Western Caribbean, from Cozumel and Roatán to the tiniest cays, are porous limestone, what some in the English-speaking part of that region call "ironshore." They function as extensions of the Mesoamerican Barrier Reef.

At the outer edge of the lagoon, the reef thrusts up sharply. From the balcony of a condo or even from the beach, the crest of a Caribbean barrier reef is obvious. This is where the big waves break, in a long, regular line. On one side, the water is deep blue; on the other, it is turquoise, infused with sunlight reflected from the sandy bottom of the lagoon. Here and there the relatively shallow lagoon water takes on a darker aquamarine hue above patches of seagrass.

The flat top of a barrier reef may be anything from a few yards wide to a quarter mile or more.[57] Small islands may even dot the crest, as they do off the coast of Belize, where the barrier reef complex ranges from half a mile to 18 miles wide.[58]

From the crest, the barrier reef slopes seaward, often in what marine geologists call a "spur-and-groove" structure,[59] in which colonies of stony corals rise on either side of chutes of sand 10 to 20 feet wide. These structures resemble small canyons, and sometimes popular dive sites have "canyon" in their names. But unlike true canyons, this spur-and-groove fore reef results not from erosion, but from the corals accreting;[60] and the sides, rather than being smooth, are marked by ledges and crannies, favored shelters for crabs, eels, and nurse sharks.

After sloping gradually to between about 45 and 66 feet or stepping down in a set of terraces culminating at around 90 to 120 feet,[61] the outer reef drops off into a sheer wall,[62] an underwater cliff. Sea fans and vase sponges, some of the latter big enough to sit in (although responsible divers no longer do that, tempting though the photo op may be), filter their food from the current flowing along the drop-off, which may descend a mile deep or more. Schools of yellow-tail snapper, yellow- and turquoise-striped jacks, and flashy silversides wave collectively. Out over the blue depths, spotted eagle rays glide, and black-tipped and tiger sharks undulate in search of schooling fish.

Along the wall and down to 450 feet lies another environment entirely, which scientists call "mesophotic."[63] Although mesophotic ecosystems represent 80 percent of reef habitat worldwide, they became a popular area of study only in the present century. The first international workshop on them took place in 2008 in Jupiter, Florida.[64]

As the water deepens, the light shifts away from the red and yellow end of the spectrum, toward the blue. Below a depth of about 165 feet, well beneath the 120-foot limit for recreational diving, hard corals no longer dominate the reef. The scattered colonies that remain are plate-shaped, spreading out to catch the scarce light. Below 300 feet, even these disappear.[65] Macroalgae cover also tends to diminish with depth. For example, off Utila in the Honduran Bay Islands, macroalgae covers 45 percent of the reef at 45 feet but only 8 percent at 120 feet.[66] The deep reef is the realm of sponges, tree-shaped black coral (family: Antipatharia), and species of fish able to withstand extreme water pressure. Remarkably, some "depth generalists" that spend most of their time on the upper reef also venture here.[67] For example, lionfish, a recent invasive and voracious

Indo-Pacific arrival on the Mesoamerican Barrier Reef, have been spotted as deep as 1,000 feet.[68]

In most parts of the Mesoamerican Barrier Reef, surf breaks across the seaward side, while the lee side provides a relatively calm environment for fish, crabs, shrimp, squid, and soft corals. Able to venture below the surge of the waves, which usually requires only about 20 feet of depth, scuba divers can float above coral heads, observing delicate feather worms, scary-looking but seldom threatening moray eels, multicolored parrotfish, and red-and-white banded shrimp. Spiny lobsters hide under ledges. A bit of patience rewards the visitor with a teeming world of diverse underwater life, all within a section of reef only a couple of yards across.

Corals die quickly but recover slowly.[69] As the polyps in a mound of coral die, under normal environmental conditions new polyps replace them, each building on the calcium carbonate exoskeleton of the previous generation. Occasionally, a storm, an earthquake or a collision with a boat or a large animal will break off a piece of living coral, tumbling it to one side, where it might begin to build another coral head.

Bleaching, however, is another matter. Unlike the natural cycle of death and growth, it signals a reef that could be in grave danger. Bleached corals are not, as media have reported erroneously in recent years, necessarily dead, although they may die of starvation if persistent high temperatures cause them to remain bleached for too long. And bleaching renders living corals vulnerable to lethal epidemics such as white band disease and stony coral tissue loss disease and to storm surges and other threats that could kill them but that healthy corals might withstand. A bleached reef is like a person with critical vitamin or mineral deficiencies.[70]

This is how bleaching works: Corals' calcium carbonate exoskeletons are white; the polyps are translucent. What gives corals their color are their Symbiodiniaceae, more loosely known as Zooxanthellae, the single-celled algae that live among them symbiotically. All hard corals depend to some degree on this symbiosis. Without it, there would be no reef.[71]

Hard, reef-building corals face a dilemma: they need clear water, but clear water delivers too little plankton to keep the polyps well-fed. The evolutionary solution has been to join forces with Symbiodiniaceae. These primitive creatures line the tubular mouths and inhabit the tissues of the polyps and provide the corals with a nutrient boost through photosynthesis, which relies on the carbon dioxide the polyps produce. To fuel this process the Symbiodiniaceae use the waste products from the polyps, making for more efficient elimination.[72] For reasons scientists have yet to figure out, when coral colonies come under extreme stress—say, from a rise in ocean temperature or a flood of pollution—the Symbiodiniaceae either stop photosynthesizing or rev up their photosynthesis, producing oxygen at levels toxic to their coral hosts. The polyps respond by expelling the friendly algae, becoming an alarmingly skeletal white,[73] like when a person carelessly spills Clorox on a colored shirt. This is why this distress signal is called bleaching.

Scientists used to think that the Symbiodiniaceae belonged to a single species of algae; but recently developed genetic technology has revealed that they are diverse, and that some species are more resistant to temperature rises than others. For many Symbiodiniaceae, a mere two-degree rise in water temperature can cause them to stop photosynthesizing. Thus deprived of food, the hard corals give them the boot—the same with Symbiodiniaceae that produce toxic amounts of oxygen. The corals "bleach," their growth slows and they will starve to death if temperatures don't fall quickly back into the range acceptable by their colorful symbiotic algae.[74]

Denied the nutrients that their Symbiodiniaceae provided and unable to move to find healthier conditions elsewhere, corals sit there on the reef, helpless against whatever natural catastrophes and human abuses hurl their way. The ability of individuals, populations, and communities to return to a desirable state after being challenged by a disaster, whether an earthquake or a major bleaching event, is known as "ecological resilience."[75] The Mesoamerican Barrier Reef possesses a certain amount of natural ecological resilience, because corals have genes that they use to defend against disease and heal damaged tissue. These genes are "turned on" in the face of stressors. Scientists can measure when this "defensome" is activated. This measurement helps them identify the stressors. Armed with this information, researchers can encourage their fellow humans to do something about these threats before it's too late to help.[76]

When stony coral tissue loss disease (SCTLD), previously unknown to science, suddenly appeared off Miami in 2014, it quickly moved north and south. The devastatingly lethal and highly contagious pathogen overwhelmed the reef-building corals' defensome.[77] By 2018 it reached the northeastern coast of the Yucatán. Within another year, it had spread south along the Mesoamerican Barrier Reef, hitting northern Belize in July 2019.[78] By December of that year it had spread to eight countries around the Caribbean, afflicting 22 species of hard corals. Colonies that were infected died within weeks. Brain corals were especially vulnerable to infection, but the toll exacted on the rare pillar coral was the most devastating, killing 90 percent of the polyps afflicted.[79] In its 2020 Report Card on the condition of the Mesoamerican Barrier Reef, the Healthy Reefs Initiative, an international consortium that monitors the health of the reef and the communities that rely on it, called SCTLD the most lethal coral disease ever encountered, because it affects such a broad range of corals and spreads so rapidly.[80]

Even if the coup de grâce for a stretch of coral reef is an act of nature, the main cause of the bleaching that weakens the corals to the point where they are defenseless against it is the warming of the oceans, where according to the 2019 report of the United Nations Intergovernmental Panel on Climate Change the temperature has risen 1°C (1.8°F) since the start of the Industrial Era[81] around 1800—and the main cause of that warming is the carbon dioxide that humans are recklessly and relentlessly pumping into them.

The earth's oceans dissolve a third of the planet's carbon dioxide, including the ever-increasing amounts produced by burning fossil fuels.[82] Microbes store much of that carbon, sending it up the food chain to the animals that consume them.[83] In photosynthesis, plants, algae, and some bacteria, on land and in the ocean, use carbon dioxide to convert energy from sunlight into oxygen and simple sugars, both of which are essential to life on earth. But that process doesn't use up all the carbon dioxide humans have been pumping into the sea. Much of the excess gets converted into carbonic acid, resulting in another threat to coral reefs: acidification.[84] That acid dissolves the calcium carbonate that constitutes the major building block that new polyps need to grow their exoskeletons and makes it more difficult for hard corals to form colonies. Plummeting pH also makes it harder for mollusks to make their shells. At the levels the oceans are acidifying, some existing calcium carbonate structures are beginning to erode.[85] If the pH of the ocean drops sufficiently, coral reefs will dissolve away.

"Temperatures have risen and fallen often during the history of the Earth, but acidity is usually much more stable. No one knows what the outcome of this giant chemistry experiment will be . . . ," marine biologist Nancy Knowlton noted in *Citizens of the Sea: Wondrous Creatures from the Census of Marine Life.*[86]

Bleached corals can recover. But to do so they need relief from the stresses that caused them to expel their symbiotic algae. As things stand now, humans are exacerbating these stresses, not just by contributing to global warming and ocean acidification but in ways that can be addressed relatively easily and quickly—untreated sewage, unsustainable fishing practices, uncontrolled resort development, and much more. Even though we may not be able to reverse climate change, or even halt it as fast as we should, we still can cease such assaults on the oceans as we work simultaneously to put the brakes on global warming. As chapters 17, 18, and 19 describe, people along the Mesoamerican Barrier Reef are showing the world how.

It took thousands of years for the Mesoamerican Barrier Reef to grow, and this growth required the interaction of thousands of different species. How horrible it would be if our one species were to wipe out this fascinating, fragile ecosystem within just a few centuries, as we may well be on our way to doing.

2

~~~~~~

WHO KNEW, AND WHEN?

Discovering the Caribbean's Greatest Barrier Reef

In February 1972 marine scientists Klaus Rützler and Amfried Antonius and five of their colleagues from the National Museum of Natural History packed into an open boat and pointed its bow toward Glover's Reef, an atoll 27 miles off the southern coast of what was then still the Crown Colony of British Honduras.[1] Ten years earlier, Rachel Carson's *The Silent Spring* had been published, shocking Americans awake to the poisoning of the environment with herbicides and pesticides and shifting nature-friendly sentiment from Teddy Roosevelt–style conservationism—preservation of scenic tracts of land and iconic plant and animal species for the enjoyment of future generations—to awareness of the interconnectedness of all life, the understanding that chemicals sprayed on marshes would make their way through the fish that fed on the mosquitos to the people who ate the fish, concentrating the toxins' strength geometrically along the way. By 1970 "ecology" had become a household word; environmentalism had become a political rallying cry, right along with civil rights, feminism, and disaffection with the Vietnam War.

Noticing the deterioration of the reefs off South Florida, the Smithsonian Institution, which had built the museum in 1910 on the National Mall in Washington, DC, decided to explore what was happening to the shallow-water marine environment in and near the Western Hemisphere's tropics. The institution set about establishing a network of observation stations. One was in Maryland, another about halfway between Miami and Jacksonville, Florida, and two in Panama, one on the Caribbean side, the other on the Pacific. The Smithsonian wanted the fifth to be off the coast of what would, in 1981, become the independent country of Belize.

As entities go, the Smithsonian Institution is an odd duck best described as "quasi-governmental." It was founded in 1846 with funds from the estate of James Smithson, a British scientist who had never set foot

Scouting location for Smithsonian Institution field station, 1972. *Left to right:* Walter Adey, Arthur Dahl, Tom Waller, Klaus Rützler, and Amfred Antonious, photo by Mary Rice for the Smithsonian Institution.

on North America. Amounting to more than half a million U.S. dollars at the time of his death in 1829, Smithson's bequest was a surprise to everyone except presumably the solicitor who drew up his will. Historians have speculated that the scientist intended the posthumous gift as a poke in the eye to England. He was the illegitimate son of a rich Englishman, and although his father had acknowledged him, the courts had ruled that he was not entitled to inherit. Whatever the reason Smithson chose the United States over his native land, he stated his broader purpose clearly: to fund "at Washington, under the name of the Smithsonian Institution, an establishment for the increase and diffusion of knowledge."[2]

It took President Andrew Jackson seven years to inform Congress of Smithson's beneficence and another eight for that cantankerous body to draw up an act establishing the Smithsonian and specifying that it be administered as a charitable trust by a board of regents and one "Secretary of the Institution."[3]

From the perspective of their offices in Washington, DC, in the early 1970s, the Museum of Natural History scientists thought that Glover's Reef would be the ideal spot for a permanent field station to participate in what came to be called the Caribbean Coastal Marine Productivity program (CARICOMP).[4] (Half a century later, CARICOMP had grown into an international network of marine laboratories, parks, and reserves studying the interaction between the land and the sea throughout the Caribbean Basin. Among CARICOMP's contributions has been the establishment of consistent protocols for collecting data and a center for processing it, at the University of the West Indies in Kingston, Jamaica.[5])

Although Glover's Reef sat in international waters, administratively it was part of British Honduras, yet far enough offshore to avoid the direct effects of runoff from the mainland (granted that this runoff might be an important factor influencing the health of the ecosystem and thus worthy of study).

Exploring the possibility that Glover's Reef might be a good site for a field station, the seven marine scientists were looking for the Tobacco Cay (pronounced "key") Cut, a natural channel through the crest of the coral, when a storm began to kick up. They were 10 miles off the coast, too far to turn back. Aware that they needed shelter to wait out the weather, they noticed nearby a little island sporting 58 coconut palms and a few frame buildings.[6] Pulling up to its dock, they disembarked and were greeted by Henry and Alice Bowman, members of the business clan who had recently built the Pelican Beach Motel, the first modern accommodation in Dangriga, a small coastal town where the economy focused on fishing and citrus.[7]

While the scientists waited for the storm to die down, they chatted with their hosts, who were curious about what they had been hoping to do on Glover's Reef. Rützler explained that they were looking for a site for a field station, part of the network the Smithsonian was setting up to monitor the health of tropical coral reefs. Because the field station would serve as a location for the study of the biology, geology, and ecology of coral reef ecosystems, it would need to offer easy access to reef environments, seagrass habitats, and mangrove islands.

Why didn't they lease the island, Carrie Bow Cay? Henry Bowman asked. It fit their criteria and already had two houses and a manager's cottage, so they wouldn't have to start from scratch. The family used the cay only on holidays and occasionally during school vacations, Bowman explained. His three siblings and their children preferred the Bowmans' compound on South Water Caye (the little islets can be spelled either with or without the final *e*); at 15 acres it was more than seven times as large. He assured Rützler and his colleagues that the family wouldn't get in the way of the researchers. Compared to Glover's Reef, Carrie Bow Cay would be much more convenient. The shops and air strip of Dangriga would be only half an hour to 45 minutes away by boat, less than half the travel time. And the cay sat right on top of what was then known as the Belize Barrier Reef.

Henry Bowman and Klaus Rützler struck a deal.[8]

As measured in a straight line, about a third of the Mesoamerican Barrier Reef lies along the coast of Belize. Culturally, Belize seems more like an island in the British West Indies than part of Central America and is unique in the region as having English as its official language. On a stay near Dangriga in 2016, I went snorkeling one day with a guide named Medel, and the previous day my scuba guide had been Ganif, first names that sounded not only English but downright Tolkienesque. The two young men belonged to the 6 percent of Belize's 402,154 estimated 2021 population[9] known as Garifunas,[10] descendants of Africans and Carib Indians whom the British had driven west from the islands of the Eastern Caribbean, presumably because they were too independent-minded to make good slaves.[11]

The Garifunas are one of Belize's smaller minorities. In the 2020 census, Creoles, who combine African and European ancestors, made up 25 percent of the population; Mestizos, descended from Mayas and Europeans, constituted 53 percent; 11 percent of Belizeans were Maya; 6 percent claimed mixed ethnic backgrounds, and the rest identified themselves as being of Chinese, East Indian, or undiluted European heritage.[12] The most distinctive minority are the Dutch and German Mennonites, who farm land in the north and west of the country, craft and export elegantly simple wooden furniture, and retain such customs as long calico dresses and white head scarfs for women.

Among family and friends, the local people around Dangriga speak a patois called Kriol, but they learn English in school. Although this English is fluent, Belizeans pronounce some words differently than North Americans do. One day in 2013, when they were on Carrie Bow Cay conducting their annual monitoring, scientists John Tschirky and Karen Koltes spotted a large buoy, maybe six feet tall, used to mark Marine Protected Areas where fishing was prohibited. The buoy had washed up on the fringe reef. They took a boat out to it, discovered that the line appeared to have been cut, towed the buoy to the cay's pier and used the field station's satellite phone to call the Belize Fisheries Station a couple of cays to the north.

"We found one of your buoys washed up on the reef," Tschirky announced.

Shortly thereafter, a Fisheries boat arrived. Aboard it were the head of station, dressed in his full official uniform, and two Fisheries rangers with machine guns. "Where's the boy?" the Fisheries officer asked.

"Out on the pier," Tschirky said.

Looking puzzled, the Belizean insisted that he hadn't seen any boy. Due to the way the local people pronounce and interpret words, he'd thought that the scientists had found a young man's body washed up on the reef.[13]

The Mesoamerican Barrier Reef ranks among the world's most diverse ecosystems. Statistics collated by the World Wildlife Fund record 65 species of hard corals, 350 species of mollusks, and 500 species of fish.[14] One study of the waters around Carrie Bow Cay found 165 species of marine algae.[15] Thanks to genetic sequencing, scientists identify more in these categories each year. Mouths of the rivers flowing into the reef are home to the largest population of manatees on the globe—between 1,000 and 1,500 of the huge marine mammals.[16]

In the mid-2010s Australian ecologist Lucie Bland of the Centre for Integrative Ecology at Deakin University in Melbourne led a team of scientists and data analysts to predict what would happen to the Mesoamerican Barrier Reef in the event of 11 scenarios of threat, using the criteria outlined by the International Union for Conservation of Nature (IUCN). Government agencies, nongovernmental organizations, researchers, and other concerned individuals from 160 countries belong to IUCN; so when it labeled the reef critically endangered, Bland and her colleagues wondered whether, and under what circumstances, it might disappear.[17]

To get an idea of the condition of the Mesoamerican Barrier Reef before humans arrived en masse on the coast and the cays, Bland's team studied the healthiest spots along the present-day reef and melded those findings with data from the 1970s. Then they took five current threats—hurricanes, overfishing, mass bleaching, ocean acidification, and pollution (from both nutrients and silt)—and created likely scenarios. For example, what would be the result if a major hurricane hit during an episode of mass bleaching?[18]

The good news, if it can be called that, was that the researchers concluded that no combination of the threats studied was likely to cause "ecosystem collapse" by 2037. The bad news was that the balance of life along the Mesoamerican Barrier Reef was precarious. Depending on the scenario, between 28.9 and 93.1 percent of the coral cover would be destroyed; the biomass (a combination of number and size of individuals) of herbivorous fish would drop between 50.2 and 82.7 percent and of the fish that prey on them, between 36.8 and 81.5 percent.[19]

The rosiest analysis marked some portions of the Mesoamerican Barrier Reef as merely "Vulnerable," but overall Bland's team concluded that the reef as a whole deserved to remain on the ICUN "Red List" as "Critically Endangered." And that scenario in which ever-stronger hurricanes hit bleached corals? That was the worst of all.[20] The destruction of 93 percent predicted to result from the dual catastrophe would leave only 7 percent of corals alive to help the reef gradually regenerate. That would come perilously close to "ecosystem collapse," the term ecologists use to describe an ecosystem so devastated that it could never restore itself.

The first scientific investigations focusing on the barrier reef and the cays were conducted as part of the Cambridge Expedition to British Honduras in 1959–1960.[21] Belize already had been conservation-minded for a generation. To protect the red-footed booby, in 1928 the then-colony declared the bird's rookery 55 miles east of Belize City on Half Moon Caye a crown reserve bird sanctuary. After Great Britain granted Belize independence on September 21, 1981, the preserve became the new country's first site for the protection of wildlife. On March 4, 1982, Belize granted national monument status to 41.5 acres of the islet and Lighthouse Reef, the atoll on which it stood, which includes the iconic Blue Hole vertical subaquatic cave.[22] Among the first laws passed by the new parliament had been the National Parks Systems Act of 1981. In 1996 UNESCO named the Belize Barrier Reef a World Heritage Site.[23]

By 2015, the Mesoamerican Barrier Reef had 45 Marine Protected Areas, where fishing and the harvesting of lobsters, conch, and other species are either forbidden altogether (except in the case of lionfish, an invasive species) or strictly controlled. They include Arrecifes de Cozumel National Park in Mexico, Hol Chan Marine Reserve, and Sian Ka'an Biosphere Reserve in Belize, and Cayos Cochinos Marine Park in Honduras. The range of protection varies from one to another.[24] Although fishing for anything but lionfish was outlawed altogether in only 3 percent of the reef ecosystem, 20 percent of it had some restrictions in place.[25]

But Marine Protected Areas are effective only when local fishers abide by the boundaries and authorities can enforce them. A few days after I arrived at Carrie Bow Cay in September 2016, John Tschirky took me over to the Belize Fisheries station, a rustic affair consisting of two pink frame buildings, an outhouse, and a small white structure incongruously sporting a sign above its lintel announcing, "Info and Gift Shop." As the ranger helped us beach our panga, a high-prowed open boat resembling a Boston Whaler, Tschirky mentioned having spotted some men that morning fishing in a "no-take" part of the channel separating Carrie Bow from South Water. Tschirky had tried to call in a report, but the station's phone had given him the message that it was out of service. The ranger explained that he'd been out to the fishing boat to sanction it, and the fishers had said that the line between the conservation and the no-take zones was marked so poorly that they couldn't tell which side they were on. Because the ranger had caught one of the men fishing in the same area several times before, he hadn't bought the story but admitted that Fisheries needed to mark the boundary better. A stack of brand-new red marker buoys were piled up near the dock, which had been ruined by a recent storm. The ranger explained that he was waiting for the hardware necessary to set them. Who knew when it would be delivered?

For increasing local buy-in for fishing regulations and in general protecting the country's share of the Mesoamerican Barrier Reef, Belize and Mexico are emphasizing education. In Quintana Roo, the Mexican nonprofit Comunidad y Biodiversidad (COBI) trains local fishers to scuba dive and to collect data on the health of the ecosystem. These trained citizen-scientists also serve as informal ambassadors for sustainable fishing practices and conservation, educating fellow members of their communities, not just informing them but instilling a sense of ownership and responsibility.[26]

Another Mexican nonprofit, Guardianes del Arrecife (Guardians of the Reef), trains locals in Puerto Morelos, an erstwhile fishing village 20 miles south of Cancún on the booming Riviera Maya, to repair corals broken by storms or boat hulls. Guardianes del Arrecife also teaches locals passive protection of the environment. Because much of what had been their fishing ground was now off-limits as Puerto Morelos Reef National Park, many of the fishers had become tour guides. Teaching them, and their clients, to avoid touching anything underwater is critical.[27] "If we teach the people the basics, that does a lot of good," explained Gustavo Guerrero Limon, a marine biologist who participated in the training.[28]

Medel, my Belize snorkeling guide and a native of Hopkins, a small village south of Dangriga, was proud that the underwater wonder where he worked was recognized as an international asset and acknowledged as the second-longest coastal barrier reef on the planet. Beginning in primary school, he and his classmates had learned about the environmental importance of what was then known as the Belize Barrier Reef. The World Wildlife Fund led the movement to consider the health of the Mesoamerican Barrier Reef a multinational issue.[29] What happens in one section of the reef or the lagoons or land abutting it can affect the entire 625-mile-long ecosystem.[30]

Initially, many Belizeans resisted the change of name and focus. "They felt that their reef was famous and that they'd lose that," explained Melanie McField, director of the Healthy Reefs for Healthy People Initiative (HRI), which she co-founded in 2004 to promote the well-being of both the Mesoamerican Barrier Reef and the human population that depends on it. But the collaboration and participation of other nongovernmental organizations with HRI increased international appreciation of the reef, in turn helping Belizeans recognize that it was a global treasure and that its multinationality could be a source of national pride.[31] Soon in every grade, Belize schoolchildren were learning about the interconnectedness of this fascinating and fragile environment.

Ironically, one of the main challenges to the health of the reef is its own beauty, which draws millions of tourists a year. The boom began in 1970. On Mexico's west coast, Acapulco had been a popular escape for North Americans for a generation, and Puerto Vallarta was attracting wealthy Californians, beginning with the cast and crew of the 1964 hit film *The Night of the Iguana*. But across the country, on Mexico's Caribbean coast, residents struggled to survive as fishermen or as laborers on coconut and chicle plantations. The eastern edge of the Yucatán, Quintana Roo, was not even a state yet. Some adventurous tourists made their way to Cozumel in the 1920s and 1930s, staying at a few hotels built primarily to accommodate American importers of chicle, the natural ingredient in chewing gum; but when chicle was replaced by a synthetic during World War II, the hostelries began to close. Thanks to magazine articles, the island's tourism received several boosts during the 1950s, as treasure divers began exploring shipwrecks in nearby waters via newly invented scuba gear. But

Cozumel and Quintana Roo more generally remained a rarely visited tropical paradise.[32] The same was even truer of British Honduras and the Caribbean coast of Guatemala and Honduras.

Inspired by Acapulco, Puerto Vallarta, and the charming colonial cities and striking pre-Columbian ruins of Mexico's interior, all of which drew international visitors as well as prosperous Mexicans, some innovative bureaucrats in Mexico City reasoned that tourism might be the answer for areas blessed with natural wonders but cursed with economic malaise. What about creating master-planned resorts, rather than just waiting for existing beach towns to be discovered? Identifying Ixtapa on the Pacific and Cancún on the Caribbean as having the greatest potential, in 1968 the Bank of Mexico floated a loan from the World Bank and set to work.[33]

In 1970 Cancún had only three residents, caretakers for a coconut plantation owned by Don José de Jesús Lima Gutiérrez, a wealthy man who lived on a nearby island, Isla Mujeres. No private investors seemed interested in funding the scheme, so the government financed Cancún's first nine hotels. In 1974 Quintana Roo became a state, and Cancún became an Integrally Planned Center, one of the initial projects of a government entity that evolved into the Fondo Nacional de Fomento al Turismo (FONATUR).[34]

That initial injection of capital paid off. Between 1970 and 2020, Quintana Roo's population skyrocketed from 88,150 to 1,723,259,[35] with 16.4 percent of the state's economy based on accommodations and food services and another 5.1 percent on arts, entertainment, and recreation.[36] The coast of Quintana Roo had become the Mayan Riviera. By 2019, 3.3 million passengers a year were disembarking at Cozumel's three cruise ship ports.[37]

And what had been pristine sections of coral reef experienced everything from carelessly flung boat anchors to heedless disposal of sewage.

Belize has undergone less of a touristic boom than Quintana Roo and is managing it better; but large-scale tourism is still a threat.

"The problem is always going to be carrying capacity," explained Therese Bowman Rath, who manages the Pelican Beach Resort, which began as the Dangriga motel built by her father, Sir Henry. "The overnight tourism is manageable. People who come specifically to Belize and stay several days or a week come for the environment and are prepared to protect it. What we worry about is over-expansion of the cruise ship industry. Whenever you have volume, you have more chance of destruction.

We also have the potential for overfishing, because those visitors have to eat."[38]

The 2020 COVID-19 pandemic brought a global halt to the cruise industry, but as soon as the Centers for Disease Control lifted its "No Sail" order that fall, companies began marketing spring 2021 itineraries for the Western Caribbean.

People who book vacations on 3,000-passenger cruise ships consider the ship and its amenities their destination. The appeal of ports of call is generally secondary—shopping, whirlwind guided tours, half-day zip line and ATV adventures, maybe a day on the beach or a couple of beers and some seafood under a *palapa*. Just injecting into the environment 3,000 happy holiday-makers and another 1,000 or more crew members (the thousands of workers on board take turns having time off in port) alters the ecosystem, not to mention the local culture.

The passengers of giant cruise ships aren't coming to Belize, Quintana Roo or the Bay Islands of Honduras. Any other lovely tropical paradise within a day or two of home would do. And if in 20 years the Mesoamerican Barrier Reef is dead and the once-limpid waters around Cozumel, Cancún, Playa del Carmen, Belize City, and Roatán have turned murky, there will always be another tropical paradise.

Maybe.

3

UNEASY SYMBIOSIS

The Relationship between the Mainland and the Reef

When we think of tropical reefs, we think of coral. And hard corals are, indeed, the ecosystem's backbone. The soft corals, sponges, mollusks, fish, and other animals that inhabit the reef are its organs. But the muscles and the skin—those are the lagoons, the mangroves, and the sea. The health of the whole is dependent on the health of each part.

Like most metaphors, this one is imperfect, and yet it conveys an important reminder: The reef, the coast, and the sand flats and seagrass beds between them are interdependent. Anything that threatens one threatens all.

Devoted to protecting and enhancing the reef and the lives of the human beings who depend on it, the Healthy Reefs for Healthy People Initiative (also known as the Healthy Reefs Initiative or HRI) identifies the Mesoamerican Barrier Reef ecosystem as covering 179,313 square miles. This intricately interdependent web of life includes 74,382 square miles on land, watershed that drains into the Caribbean Sea. The remaining 104,931 square miles lies in the Caribbean—in the lagoons and the waters covering or on either side of the coral reef.[1] Apart from its exquisite aesthetic qualities, this ecosystem is very valuable economically, especially for tourism and commercial fishing. In Belize, the reef contributes between US$395 million and US$559 million annually to the gross national product.[2]

Scientists used to think that marine species were replenished by distant populations. Many marine animals do travel great distances during parts of their lives, and the oceans' inhabitants are much more likely to be interconnected genetically than plants and animals on land. However, even some world travelers mate or spawn close to where they were born or hatched, so local ecosystem management is critical.[3]

Big marine animals (tuna, sea turtles, whales) often migrate thousands of miles to spawn or mate, making protection of breeding habitats particularly important.[4] And protection of breeding habitats depends on effective enforcement at the community level.

Especially for those species that gather in specific areas to breed, destruction of the local environment poses the danger of extinction. No one knows how many species of plants and animals inhabit the earth; with the aid of DNA sequencing technology, scientists are identifying new ones constantly. Granted, species disappear due to natural causes ranging from lethal diseases to erupting volcanoes, but human activity multiplies the rate of such "background extinction" manyfold. According to conservative estimates, humans are responsible for the loss of 200 other species a year.[5] Even species that survive our carelessness and become merely rare may be genetically damaged due to inbreeding.[6]

Traditionally, the least valued part of the Mesoamerican Barrier Reef ecosystem has been the mangroves that cluster along the shore and form little islets on top of submerged patch reefs. Investors see the short, densely massed trees with their roots sunk in the water as a barrier to be cleared away and replaced by sand trucked in for beaches. The occasional birder or other amateur naturalist might enjoy kayaking around a clump of mangroves, focusing a pair of pricey Austrian binoculars on a graceful white egret perched on a branch or a green iguana basking in the dappled sunlight; but to most visitors, mangroves are a turnoff. You can't lounge on them. You can't wade off them.

To make way for farms and resorts, more than a third of the world's mangroves were ripped out between 1990 and 2010. What a loss! Mangroves not only protect shorelines from waves; as chapter 1 explained, they also provide nurseries for fishes,[7] juvenile crustaceans, and baby turtles. Seventy percent of the Caribbean's sports fishing species and 90 percent of commercial species spend some part of their life cycle among the mangrove roots.[8]

"Take the parrotfish," said marine biologist Gustavo Guerrero Limon, who served as a volunteer with Mexico-based nonprofit Guardianes del Arrecife in Puerto Morelos, Quintana Roo. "The larvae have to move back to the mangroves until they get big enough to move to the reef. Hard corals, seagrass beds, mangroves—these three ecosystems cannot be separated. They cannot be split."[9]

Forest allowed to reach down to the lagoon, photo by Emily Kasten, courtesy of Hamanasi Resort.

Above the surface, the mainland swamp across from the Mesoameri-can Barrier Reef is one of the least diverse tree canopies on the planet. Because red mangroves tolerate salt better than the white and black varieties,[10] they are one of only a few species of trees that grow there.[11]

And beneath the surface, this same arboreal monoculture holds the future of the ecosystem in its roots. "Mangroves are a nursery ground for species that will spend their adult lives offshore," Texas A&M University at Galveston marine biologist Gilbert Rowe explained. "Mangroves export material from the land to the reefs, because their leaves fall off and decay in the water, which washes out to the reefs."[12]

While still attached to the tree, mangrove leaves shade shallow-water creatures and birds' nests from the intense tropical sun. Brittle stars and boxy starfish wrap themselves around mangrove roots next to soft corals, oysters, sponges, and tunicates, which are small filter feeders.[13] Bacteria transform leaves and barks shed by the mangroves into mulch, food for little worms and crustaceans, which become food for small fishes, which are then eaten by larger fishes, which in their turn become meals for wading birds, birds of prey, crocodiles[14], and people.

This subsurface environment may even hold the potential to save human lives: one of the creatures that live among the roots, the sea squirt, produces a chemical with anticancer properties.[15]

Henry Bowman experienced a vivid demonstration of the value of mangroves in preventing erosion. To create a beach that his family could enjoy at their vacation home, in 1944 he replaced the mangroves on Carrie Bow Cay with coconut palms. Although their roots helped hold the sand in place, the palms offered scant protection from storm surges.[16] By 2020, the islet had shrunk from two acres to less than three-quarters of an acre.[17]

Texas A&M University at Galveston marine biologist Tom Iliffe told me how the destruction of the mangroves works on a larger scale along Mexico's Rivera Maya, where the law prohibits clearing mangroves along the shore but permits "improving" land for a golf course or other resort amenity a ways back from the water.

"As the new development builds up, they close off the connection between the mangroves and the ocean," Iliffe told me. "The developers clear off the mangroves except for a little fringe, and they clear out the area between the mangrove fringe and the beach. Then at five o'clock on Friday afternoon, they bring in the bulldozers and clear out the fringe, which they have the whole weekend to do.

"It's much simpler to ask for forgiveness than for permission," he added. "The developers pay the $10,000 fine, but they now have property worth $10 million."[18]

Meanwhile, the environment has lost an asset worth incalculably more. In addition to the local protection they provide and to their role as nurseries for aquatic species, mangroves are up to six times more effective at capturing carbon than rain forests are, so good that mangrove planting and maintenance can now be traded on the international carbon market.[19] Mangroves help filter out the silt in runoff. Following a heavy rain, they slow the introduction of fresh water into the saltwater environment, promoting the growth of offshore coral reefs. The process isn't just physical, it's also biochemical.[20]

Between the mangroves and the coral lie saltwater lagoons, which bustle with diverse life. In the shallow waters just off the mangroves, meadows of seagrasses spread their blades to photosynthesize. With about 70 different species,[21] seagrasses populate lagoons and bays in all but the world's coldest oceans.[22]

All seagrasses evolved from flowering land plants that returned to the ocean. Somehow they adapted themselves to be able to pollinate underwater.[23] In addition to spreading via the resultant fertilized fruit, seagrasses also reproduce asexually by sending out rhizomes, much like the Saint Augustine and Bermuda grasses that grow in North American lawns. Half of a seagrass plant extends below the lagoon floor.[24]

Certain algae grow on the grass blades, as do bacteria and films that feed juvenile fishes and invertebrates. Some fishes like to munch the grasses themselves. Turtles crop them like a lawn mower, but manatees often tear them up by the roots. Common in the Mesoamerican Barrier Reef ecosystem, shoal grass (*Halodule wrightii*) has flat, slender blades; manatee grass (*Syringodium filiforme*) grows long, spaghetti-like ones. The most common seagrass in the Caribbean, turtle grass (*Thalassia testudium*), sports erect green blades up to two feet or more long; its rhizomes trap silt, cleaning the water for the clarity demanded by hard corals. In trade, the reef gives the grasses the protection they need from surf and currents.[25]

But seagrasses can do only so much cleansing. Heavier rains from increasingly severe tropical storms have combined with poor agricultural and development practices to wash more silt into the system, where it settles, waiting to be churned up the next time the sea gets rough. "The water is never as clear as it was, because a smaller event, like a squall, can stir the silt up from the bottom and the seagrass," John Tschirky explained.[26]

In two decades of monitoring water clarity at Carrie Bow Cay, he and Karen Koltes have documented a constant downward trend, bad news for reef-building corals.[27] Protecting seagrass meadows and encouraging their growth would, in turn, help improve conditions for the entire Mesoamerican Barrier Reef. (Mexico has enacted regulations protecting seagrasses.[28])

To photosynthesize, seagrasses depend both on such waterborne

Turtle grass bed with sea egg urchin, photo by Alex Poli.

detritus as shed mangrove leaves and on phytoplankton. Their roots can sequester carbon from these sources.[29] Seagrass beds constitute important carbon sinks not just locally (e.g., for the Mesoamerican Barrier Reef) but also globally.[30]

The first official recognition of the ecological importance of these aquatic plants came in the form of the 1973 International Seagrass Workshop held in Leiden, the Netherlands. Based on a synopsis of existing literature at the time, the participating scientists concluded that seagrass beds were among the most productive and complex of oceanic ecosystems. Most importantly, they recognized that seagrasses were often just one component of a highly diverse ecosystem.[31] The seagrass ecosystem is especially sensitive to pollution, which has increased due to runoff from poor land use practices. The resulting decline in seagrass biomass has been one cause of a decline in fisheries, because commercially valuable fishes rely on seagrasses for food either directly or indirectly, by eating creatures that eat the seagrasses and the things that grow on them. In their early phases of life, many fishes and invertebrates also rely on seagrass habitat for protection from predators.[32]

As the water in a lagoon deepens, seagrass meadows give way to sand flats. These seem almost eerily vacant—until a southern stingray (*Hypanus americanus*) flaps its way up from the bottom where it has been hiding. Scattered here and there are queen conch (*Strombus gigas, Lobatus gigas* and *Aliger gigas*), the muscular mollusks whose pink-lined spiral shells children hold to their ears to "hear the ocean." Even the conch shells tossed overboard by fishers who pierce their apex to release the tasty meat have a function, providing homes for hermit crabs and other creatures seeking sturdy shelter. (Along the Mesoamerican Barrier Reef, queen conch are now partially protected in Honduras, Mexico, and Belize.[33])

Gradually, small patches of hard and soft corals appear, sometimes erecting themselves on chunks of coral broken from the reef proper by storm surges. Burrowing worms, crustaceans, and jawfish emerge from under these. Rays glide by. Squads of squid hover. And then the barrier reef looms, building toward its hard coral crest.

Some reef-building corals rank among the most endangered animals on the planet, a list that includes a third of all coral species,[34] many of them native to the Caribbean. The Mesoamerican Barrier Reef is bathed by the Caribbean Current, which sweeps in from the east across the Bay Islands of Honduras, then north along the short stretch of Guatemalan coast and on past Belize and Mexico's Yucatán. The current takes pollution and runoff—in short, all the results of faulty land management—from south to north.[35] It also transports algae spores, seeds such as coconuts, and newly spawned animals, including coral larvae.

The hard corals that form the foundations of reefs begin life as tiny larvae, called planulae.[36] Not able to swim on their own, these juvenile creatures float on currents, sometimes for hundreds of miles. When they encounter a hard surface in warm, relatively shallow water of the proper pH (acidity), they settle down to form cup-like calcium carbonate exoskeletons and develop into polyps, column-shaped animals with tentacles surrounding their mouths. The polyp separates its interior with thin calcium carbonate walls. Varying in number from species to species, these septa produce the starlike patterns that decorate the surface of dead corals.[37]

Hard corals all belong to the phylum Cnidaria, which also includes jellyfishes and other aquatic creatures that use stinging cells, cnidocytes, to capture their prey. But unlike their mobile jellyfish kin, once they reach adulthood, corals are stuck in place.

Every hard coral polyp sports six or more tentacles (but never eight), each armed with cells that can shoot out threads or whips tipped with venom, which stun prey and can result in painful, festering abrasions for humans who bump against them. Most species feed at night, brushing their tentacles through the water to snag planktonic organisms, some of them larvae of fishes, invertebrates or corals, unable to propel themselves to safety. Although dependent on the minute animals that float by, some species of corals can be highly efficient carnivores.[38] Their polyps' tentacles sweep the prey into the slit-like mouths; then enzymes in the esophagus dissolve it. Corals in a colony share a digestive system, so the nutrients harvested by one polyp are shared by all.[39]

Still, without the photosynthetic boost provided by their Symbiodiniaceae (popularly called Zooxanthellae), the polyps would eventually starve. As good as they may be at snagging prey, hard corals need more than they can catch.

Being stationary also presents a reproductive challenge. Because corals can't move around in search of mates, each species synchronizes its spawning time precisely. For example, plate corals (family: Fungiiadae) in the Caribbean release their eggs and sperm 90 minutes after sunset five days after a full moon in August or September. (The exact month depends on water temperature hitting an ideal 27°C to 28°C.) This coordination makes it more likely that the eggs of one colony will be fertilized by the sperm of a neighboring colony.[40]

Corals invest months in developing gametes. The payoff requires precise timing. Even missing by a few minutes leaves a coral colony out of the reproductive game.[41]

Success depends on complex chemistry. Light-sensing proteins respond to the changing colors and intensity of twilight. Other chemical triggers pick up on pheromones released by neighboring colonies. The synchronized release of sperm and eggs also depends on surface temperature, wind patterns, tide cycles, atmospheric pressure, and lack of rainfall.[42]

To a diver, coral spawning may look like a pink blizzard, but the process is so delicate that light pollution from a nearby resort can disrupt it. So can thermal stress from warming waters or endocrine disruption from pesticide runoff.[43]

The hard coral polyps that result from successful fertilization spend their lives growing. As they add to the length and width of their exoskeletons, the polyps divide and bud to keep pace. The top of the polyp reaches out to its neighbors to form an uninterrupted mat of transparent living tissue covered by a protective layer of mucus.[44] In some species of corals, the polyps space themselves widely, with plenty of calcium carbonate construction separating them from their neighbors. Other species share common walls and form common valleys and ridges.[45]

Grouping together with others of their particular kind, the polyps establish colonies of different shapes, depending on their species. Some coral colonies look like church spires, others like giant mushrooms, still others like deer antlers. In terms of sheer mass, the most common along the Mesoamerican Barrier Reef are boulder corals (*Orbicella annularis*), which as their name implies form rounded irregularly shaped rocks, some over six feet high and even larger in diameter.[46] One distinctive-looking variety of boulder coral, starlet coral (family: Siderastreidae), forms hemispheres, sometimes big, sometimes as small as a golf ball cut in half. Its cups resemble black dots on buff-colored backgrounds.[47]

Because the fringed polyps attract brown or green Symbiodiniaceae, healthy boulder corals display those tones, although sand, small shells, and other detritus may accumulate in a colony's crevices, leaving white scars. To mend such damage, boulder and other hard corals can produce their own calcium carbonate cement.[48] Healthy corals can also use tiny cilia on their outer surface to shed sediments.

For divers and snorkelers, brain corals (families: *Mussidae* and *Merulinidae*) are the attention grabbers among the hard or "stony" corals. The surface of each rounded mound is a maze of convoluted ridges and valleys, like a terraced field or the surface of the human brain. In one species, the sharp-hilled brain coral, the "valleys" are a startlingly bright green.[49]

Sometimes anchored on rocks, sometimes on the calcium carbonate skeletons of defunct boulder or brain coral colonies, pillar (*Dendrogyra cylindricus*), lettuce (family: Agariciidae), elkhorn (*Acropora palmata*), and staghorn (*Acropora cervicornis*) corals reach up from the reef. These also are hard corals, with prominent polyps merged into inflexible masses, but their comparatively fragile structures render them vulnerable to damage from storm surges and boat propellers. This is especially true of elkhorn coral, which is characterized by top-heavy plates balanced on narrow stems. Only a few decades ago, this species was ubiquitous along the Mesoamerican Barrier Reef; but repeated assaults by diseases and major hurricanes have earned it a place on the "Endangered" list.

If any coral can be called "cute," it's finger coral (*Porites porites*), whose colonies insinuate themselves around and between their larger cousins. Unlike polyps of most other species of hard corals, those of finger coral routinely reach out from their calcium carbonate cups to snag food by daylight.[50] The result is a display of fuzzy-looking branches, sometimes in vivid shades of lavender or yellow, jaunty against their sober brown and olive green background.

Any coral can leave an incautious snorkeler or diver with an irritating scrape rendered worse by the venom that the polyps use to stun prey; but fire corals (family: Millepora) pack a painful wallop that can sting or itch for weeks. Marine biologists call fire corals "false corals," because their polyps reside deep inside the colony's surface.[51] This smoothness is a warning, as is the white band along the rippling edge.

Cohabiting in a healthy coral ecosystem, these varied species collaborate to form what look like fairytale rock gardens or multi-hued miniature mountain ranges. Each square yard of the barrier reef seems to contain its own world. Entire communities of crabs, shrimp, and fishes live amid the branching corals, which protect them from predators. Meanwhile, these residents drive off polyp-munching snails, fertilize the coral polyps with their excrement and even keep the coral swept clean of debris.[52]

And that is just a single example of the symphony of symbiosis. Except for a few invasive species, including human beings, this fascinating, fragile environment functions harmoniously. That harmony is the sum of the natural functions of the various species that the following section explores.

II

NATURE'S AQUARIUM

4

MASTERS OF THE AQUATIC COMMUTE

Sea Turtles

The second morning of my week as a guest observer at the Smithsonian Institution's Carrie Bow Cay Field Station off the coast of Belize, I awoke to a shout of "Turtle!" Although my watch read 4:30, it was daylight. (Belize is due south of Mobile, which lies in the Central Time Zone; but during much of the year it's on the same time as Denver, Mountain Time, and therefore an hour earlier.) Bolting out the door of my room in my nightgown, I saw Joanna Walczak, one of the assistant scientists, pointing to a loggerhead turtle with a dark, almost black, shell a yard wide. The creature was lumbering toward the beach between our no-frills shared cabin and the elevated, relatively cushy quarters called the Honeymoon Cottage. Clearly the loggerhead was a female looking for a place to lay her eggs. Although Carrie Bow Cay measured only seven-tenths of an acre, it had five turtle nests. From the looks of the distinctive tracks in the sand—marks from the tips of four flippers, with a straight line from the tail down the middle—the turtle had been walking around in circles, like a human mom plodding along a hospital corridor, hoping to encourage her contractions to speed up.

As the turtle veered toward the open storage space under the Honeymoon Cottage, the only private unit in the compound, Walczak signaled me to back off, explaining, "I think she wants some privacy."

Although the turtle looked large to me, that loggerhead was typical of her species. An adult's slightly elongated shell, called a "carapace," can reach almost four feet long. Five bony plates, or scutes, run down the middle of a loggerhead's back, with five more scutes on either side, plus one at the rear. A rim of smaller plates encircles this mosaic. Each of a loggerhead's flippers has two claws, making this particular turtle adept at digging, as well as swimming. Its most obvious distinguishing feature, however, is its oversize head armed with powerful jaws that can easily crush crabs and other hard-shelled crustaceans and even make quick work of conch.[1]

Loggerheads (*Caretta caretta*) nest in Mexico, Belize, Guatemala, and Honduras, as well as elsewhere in the Caribbean and in South Florida.[2] The loggerhead mom I saw may well have been born on Carrie Bow Cay. Like most other sea turtles, female loggerheads return to their natal beaches when they are ready to lay their eggs. Meanwhile, they travel thousands of miles, riding the clockwise gyre that circles the huge patch of the North Atlantic called the Sargasso Sea due to the large quantities of sargassum seaweed floating there.[3] In the process, turtles provide transportation for other marine life. Thirty species of barnacles live only on turtles; at least one hitches its rides solely on loggerheads.[4]

On their long migrations, loggerheads navigate using the earth's magnetic field. This enables them to avoid being swept into lethally cold waters by the northward flowing current off Portugal and helps them to find their way safely home to the Caribbean, where scent helps guide them to their natal beaches.[5]

Since the 1960s, scientists have tracked sea turtle migrations via flipper tags and more recently by satellite transmitters attached to their carapaces,[6] which is how we know how remarkably far marine turtles range. Scientists have long wondered where juvenile turtles go during the decades between hatching and sexual maturity. Thanks to powerful new transmitters and to receivers on the International Space Station, the International Cooperation for Animal Research Using Space (ICARUS), is finding out. Kate Mansfield, a researcher with the University of Central Florida, has attached satellite transmitters to turtle carapaces just five inches long. (To avoid interfering with the animals' swimming, foraging, and evading predators, the transmitters are light and strategically placed.)[7]

Of the world's seven species of marine turtles, loggerheads are among the three that nest on the cays along the Mesoamerican Barrier Reef and on the adjoining mainland's beaches. The other two are green (*Chelonia mydas*) and hawksbill (*Eretmochelys imbricata*) turtles. Leatherback turtles (*Dermochelys coriacea*) cruise for food along the reef but nest elsewhere in the Caribbean.[8]

One of the largest marine turtle nesting sites in the world is Tortuguero on the Caribbean coast of Costa Rica, where about 22,500 female

Green turtle, photo by Ben Phillips.

green turtles nest every year.[9] Green turtles are slightly smaller than loggerheads. Their oval carapaces seldom measure much more than three feet in length. Three rows of four scutes run down their backs, with an additional scute fore and aft and an encircling rim of smaller ones. The natural color of this turtle's shell is brown, but this grazer's favorite food, seagrasses, imparts a green hue to the fat underneath, as well as to the flippers and head. Green turtles bear two distinctive plates between their eyes. Only their back flippers have claws.[10]

With wavy, mottled patterns, green turtle shells are handsome; but they don't match the beauty of the hawksbill's carapace, which is composed of overlapping scutes of rich reddish brown, orange, and warm gold. During the nineteenth century in Europe and the Americas and almost to the end of the twentieth in Asia, the shell was so prized for jewelry and decorative items that the hawksbill still hovers near extinction. Today's population of hawksbills in the Caribbean is at most 10 percent of what it was before Columbus.[11] Hawksbill nesting on Mexico's Yucatán Peninsula, where 25 to 30 percent of the species' Caribbean nesting grounds are located,[12] continues to decline. In 2004, there were only 37 percent of the Yucatán hawksbill nests that there were in 1999.[13]

The hawksbill has claws on its front flippers but not the back. The turtle's sharp beak enables it to nibble sponges and pry other prey from niches in the reef.[14]

As the only survivors of both their genus, *Dermochelys,* and their family, Dermochelyidae, leatherbacks are twice removed evolutionarily from the other six sea turtle species, which all belong to the family Chelonidae. With adult males weighing in at as much as 2,015 pounds, leatherbacks are the largest marine turtles. In lieu of hard carapaces, they sport a series of bony plates covered with a layer of oil topped by one of tough skin— hence their name.[15] Along with the leatherback's breastplate, this back is flexible, allowing it to compress in response to increasing water pressure when the animal dives. In addition, their skin absorbs nitrogen, protecting leatherbacks from decompression sickness ("the bends"). Leatherback sea turtles dive as deep as 3,000 feet in search of the jellyfish that form the staple of their diet. Hard-shelled marine turtles stick to relatively shallow waters, seldom being observed below 50 feet.[16]

Other air-breathing marine animals exhale before diving, emptying their lungs to help themselves sink. Leatherbacks don't do this. Instead, they take a deep breath, apparently gauged to the depth for which they're headed. The amount of air in their lungs determines how deep they dive.[17]

Although they sometimes return to the beaches where they were born, leatherbacks aren't as faithful to these natal nesting sites as other marine turtle species.[18] They may even nest in several different countries during one reproductive season.[19] Satellite tracking has shown that leatherbacks also seem to swim farther than hard-shelled sea turtles to forage and that they tolerate the most diverse environments. Leatherbacks hatched in the Caribbean have been spotted as far north as Cape Breton in the Canadian Maritimes and even across the Atlantic off northern Europe.[20]

Marine turtles form a vital link between the Mesoamerican Barrier Reef and the mainland. They are masters of the aquatic commute, both their long distance migrations through the world's oceans and their everyday cruises from the reef to the mainland and back. Loggerheads, hawksbills, and leatherbacks are omnivorous, so they help keep crustacean, sponge, and jellyfish populations in check. Turtles also indirectly stabilize nesting beaches, where eggs that don't hatch, and even the shells of those that do, provide nutrients to the plants that hold the sand in place.[21]

Green turtles mow the lagoons between the coral and the mangroves. Their effect is generally benign, keeping the seagrass beds healthy and encouraging the plants to spread, providing more habitat for juvenile fish and crustaceans.[22] However, when sharks fail to keep the green turtle population in check, the turtles can devastate seagrass meadows by over-grazing, presenting a prime example of the ecological importance of the balance between predators and prey.[23] Here's how it works: Humans overfish the sharks. With too few sharks eating turtles, the turtle population expands, and the turtles consume too much seagrass. The seagrass meadows decline, depriving juvenile fish and larval crustaceans of protective habitat; so the numbers of fish, shrimp, lobsters, and many other species drop.

Male sea turtles can mate every year. Females wait two to five years before giving reproduction another go, although they may lay several clutches of eggs when they do.[24] Turtles may mate in the waters just off their nesting beaches, at distant foraging grounds or somewhere along the way. Then, in the course of several months during spring and early summer, the females come ashore every week or two, use their rear flippers to dig a hole in the sand and deposit a clutch of about 100 rubbery eggs.[25] Two months later those eggs that are viable hatch simultaneously, and the baby turtles work together to dig themselves out.[26] Once above ground they make a frenzied dash across the beach to the water. A little more than half make it.[27]

Kemp's ridleys (*Lepidochelys kempii*), the smallest marine turtles, hatch during the day, making the babies vulnerable to seagulls, frigate birds, and other aerial predators during their frantic sprint across the sand and their first swim through the waves. (To get the idea, take a look at the gruesome scene witnessed by Elizabeth Taylor's character early in the 1959 film *Suddenly, Last Summer.*) Baby green turtles, loggerheads, hawksbills, and leatherbacks usually hatch at night, which limits predatory threats to night-hunting animals but has made the hatchlings vulnerable lately to an inadvertent human complication: Imprinted to head for the brightest horizon, normally the water, baby turtles can become confused by lighting along the beach.[28]

Buoyed by ocean currents that convey them to the foraging grounds, the little turtles that escape being picked off by predators then spend the next several decades at sea.[29] Depending on the species, they cruise around for 25 to 35 years before they reach sexual maturity and begin the process all over again. Out of a thousand eggs laid, only one survives to reproduce.[30]

When the mother deposits them in her nest, turtle eggs have no sexual identity—they develop as one sex or the other as they incubate. The warmer the sand around their nests, the higher the percentage of females. The dividing line is a balmy 86°F. Below that, the majority of eggs in the nest become male; above that, the majority become female.[31] On the same nesting beach, turtle nests shaded by vegetation produce more male hatchlings; those exposed to the blazing tropical sun, more females.[32]

The implication of climate change is clear: As temperatures rise at the turtles' nesting grounds, male hatchlings will become scarcer. That could mean that fewer eggs will be fertilized and that those that are will be less genetically diverse.

Turtles are reptiles. All turtles, from the loggerheads that cruise the world's oceans to the box turtles children keep as pets, evolved from creatures that crawled from the ocean onto the land. Later, some of their descendants went back to the sea, perhaps because food was so abundant there.[33]

These new species evolved to fit into the marine environment. For example, they abandoned the ability to tuck their heads and limbs into their shells,[34] and they developed glands to secrete the excess salt that concentrated in their bodies as the turtles drank seawater and consumed fellow marine creatures.[35] The species that returned to the ocean did retain some of their terrestrial habits. Like us, sea turtles breathe air. But unlike us, they can hold their breaths for hours at a time. In 2020 the human record for breath-holding was 22 minutes.[36] Partly because they have much slower metabolisms,[37] turtles can remain submerged for more than an hour when they're active, four to seven when they're asleep.[38]

That doesn't mean that turtles can't drown. They can and do. One of the greatest causes of turtle deaths is snaring in fishing nets. The struggle to free themselves quickly exhausts the turtles' air reserves.

The earliest humans settling along the coasts and cays of the Caribbean relied on turtles for meat, eggs, tools, and household implements. Before the arrival of Europeans in the late 1400s, the human population remained small enough that the sea turtles held their own, although archeological evidence indicates that Indigenous hunters killed off many terrestrial animals and drove such larger sea mammals as manatees into decline.[39] Hundreds of years ago, the Caribbean had an estimated 90 million green turtles, so many that ships' captains complained in their logs that the animals posed a hazard to navigation. American colonists' appetite for turtle eggs and flesh and the popularity of items made from their shells and skin soon solved that problem.[40] Because turtles were air-breathing, they could be shipped live to England and Europe, then slaughtered for fresh meat.[41] During the 1860s, British Honduras (now Belize) annually shipped between 5,000 and 6,000 live green turtles to Britain, where their flesh was prized for green turtle soup. Populations plummeted.[42] By the 1890s, the number of green turtles exported from the colony had dropped to 50 to 150 a year, although the trade in hawksbill shell grew as the Victorians made tortoiseshell jewelry and decorative items all the rage.[43] By the 1980s several of the world's seven species of marine turtles seemed headed for extinction.

Because juvenile and male marine turtles never come ashore and they, along with females, migrate long distances, a census of individuals would be impossible; so sea turtle populations are extrapolated from nesting sites. Using this method, researchers estimate that the number of green turtles has declined between 48 and 67 percent in the last 120 to 140 years,[44] mostly *after* the drop noted in the 1890s.

This crash is due almost exclusively to human activity. Protected by their hard carapaces, adult marine turtles have few predators, except for sharks and people.[45] Scientists have yet to determine their exact life span but think that at least one species, green turtles, can live 60 to 70 years[46] if they aren't killed or don't die from complications of fibropapillomatosis, to which green turtles are especially vulnerable. The disease causes tumors on the animal's exposed skin, as well as on internal organs. These tumors can interfere with feeding if they are on or near the mouth or with spotting predators or prey if they are near the eyes.[47]

Protecting sea turtle nesting sites from human disruption does work, however. Between 2002 and 2012 scientists from the Florida Fish and Wildlife Conservation Commission monitored 16 protected nesting sites along the state's coasts, counting the number of eggs that produced hatchlings that survived marauding raccoons and other wild animals long enough to emerge from the nests. In the course of 11 years, the researchers noted a 10 percent decline among loggerheads and a 20 percent decline among green turtles, but about 60 percent of the eggs hatched successfully. Even more encouraging, the percentage for leatherback turtles almost doubled—from 22.5 percent to 41.2 percent. [48]

In 2016 during four days of diving on Glover's Reef off the coast of Belize, I saw a dozen hawksbill turtles, five at one site. Researchers might call this anecdotal evidence, but it did support the contention made by Anne Braütigam and Karen L. Eckert in their 519-page 2007 report to TRAF-FIC: The Wildlife Trade Monitoring Network and to the CITES (the Convention on International Trade in Endangered Species of Wild Fauna and Flora)[49] Secretariat that when it came to protecting sea turtles, Belize was one country in the Caribbean Basin that was doing things right.[50]

Belize has managed to write and enforce its regulations in a manner that allows traditional Indigenous use and at the same time effectively supports the sustainability of sea turtle populations.[51] In 1991, just a decade after independence from Britain, the country revised its fisheries laws to completely protect hawksbill turtles and to outlaw the taking of large juveniles and breeding-size adults of the other species.[52] Previous restrictions had focused on *minimum* sizes, based on a principle that fishers had long understood: Let the little ones go so that they can grow up to be big ones. Laws based on *maximum* size required explaining that sparing the larger turtles was essential to the preservation of the species, because the bigger animals were the ones that produced the most young.

Then in 2002, Belize enacted legislation placing strict limitations on harvesting all species of marine turtles, with the exception of a small number of individuals allowed to be taken, by permit, for traditional and cultural uses.[53] For example, the Garifuna people of the southern Belize mainland traditionally use turtle as part of the *dugu*, a ceremony honoring relatives who have died.[54]

Instead of being based on the animal's weight, as they were originally, Belize size restrictions are now based on the length of its carapace, which can be measured easily on a boat.[55] Shrimp fishers trawling Belize waters are required to outfit their nets with turtle exclusion devices (TEDs), so that sea turtles swept up accidentally can escape.[56] Belize has even implemented lighting ordinances to prevent hatchlings from becoming disoriented as they try to find the water. And Belize bans the taking of turtles by anyone, Indigenous or not, during breeding season.[57]

However, as sea turtles travel, they have no way of knowing when they cross the border from a country in which they are rigorously protected to one where regulations regarding their well-being are either lacking or loosely enforced.[58] Anne Braütigam and Karen Eckerd described what they called "a patchwork of national management regimes" among the 26 countries, territories and colonies they studied in the Caribbean Basin.[59] Half the jurisdictions lacked regulations fully protecting sea turtles.[60]

In some that had laws on the books, these were rigidly and effectively enforced. In others, they were in place but enforcement was lax, or the regulations were incomplete or outdated.[61] Unlike Belize, many of the jurisdictions exempted traditional use but were vague about what that covered; many allowed "subsistence" use but left that term open to interpretation, so that virtually any person on a cay or along the coast who fed his or her family by fishing could be permitted to take turtles, even if their meat and shells ended up in the market.[62]

South along the Mesoamerican Barrier Reef in Honduras, turtles have been a significant source of food since pre-Columbian times. Nowadays, although the country participates in international efforts to protect sea turtles, their harvesting is only loosely regulated.[63] Use by Indigenous people along the Mosquito Coast and on the Bay Islands and surrounding cays is exempted. Curios and jewelry made from hawksbill shells are sold openly, even though trade in them is illegal.[64] Factories produce body lotion and face cream made from turtle oil.[65] Sea turtle eggs are thought to have aphrodisiac properties and are sold in cantinas, in seafood restaurants and at soccer stadiums, often mixed with chili, onion, and salsa.[66]

In Guatemala, mestizos, people of mixed Indigenous and European ancestry, make up more than half the population, but more than 40 percent of Guatemalans are purely Indigenous (overwhelmingly Maya), and almost 80 percent of those live in poverty, with half the children under five chronically malnourished.[67] Turtle meat isn't popular in Guatemala, perhaps because green turtles, the tastiest species, are uncommon;[68] so the greatest challenge to the survival of sea turtles is the local appetite for their eggs, which are consumed raw, mixed with orange juice, as a morning pick-me-up either at home or purchased from street vendors.[69] After some observers estimated that every turtle egg laid on Guatemalan beaches was harvested for human consumption,[70] authorities initiated a deal: If a person looting a turtle nest contributed 12 percent of the eggs to a hatchery, he or she would get a receipt declaring the other 88 percent legal.[71]

This compromise between the needs of sea turtles and the needs of people may have been well-intentioned, but it didn't work well enough to make up for the depredation caused by the popularity of turtle eggs.[72] Exact figures are lacking, but one researcher estimated that only half the egg collectors were complying with the law, and hatcheries report that only 5 percent of the donated eggs produce viable hatchlings.[73]

Even in Belize, challenges persist. In 2020 the country finally banned the use of gill nets,[74] which inadvertently trap and drown sea turtles.[75] (Gill nets remain legal in some U.S. waters, including Puerto Rico, but not in Florida.[76]) As 2019 drew to a close, government officials had expressed a willingness to ban this indiscriminate form of fishing—but only once fishers currently using gill nets had found alternative sources of livelihood.[77] However, even in Marine Protected Areas education and enforcement remain issues. Due to tight budgets, Fisheries Department agents are spread thin along the coast and often lack money to fuel their patrol boats.[78]

On the other hand, Belize seems to be doing a good job of protecting sea turtle nesting sites. Thanks to collaborative management programs involving government agencies, nongovernmental organizations, and local communities, the number of hawksbill nests at Manatee Bar, on the mainland 20 miles south of Belize City, are increasing,[79] and recovery appears underway on the cays in southern Belize. Green turtles, loggerheads, and hawksbills all nest at protected beaches at Half Moon Cay, Turneffe Atoll, and Ambergris Cay, a prime destination for divers.[80] Because sea turtles require so long to mature, it will take three decades for efforts to pay off. But thanks to past international efforts to protect both the turtles themselves and their nesting sites, their numbers have begun to increase by between 4 and 14 percent worldwide.[81]

Yet even as we join forces to protect the world's remaining nesting beaches, we continue to kill, albeit inadvertently, the marine turtles that hatch—by producing, using, and discarding plastics. To a turtle, a single-use water bottle, a clear plastic bag, and a sheet of shrink-wrap look a lot like jellyfish, the favorite food of some species, a child's plate or occasional snack of others. And that mistaken identity can be fatal.

5

THE LIFE TRANSLUCENT

Jellyfishes and Their Kin

In my early scuba diving days, I wore a shorty wetsuit, an extreme version that covered only the top four inches of my arms and hardly more of my legs than my bathing suit. Emerging from a dive off Cozumel, I noticed a spiral welt wrapping my right leg from the top front of my thigh around behind my knee, then back to my shin, ending where my neoprene bootie protected my ankle. The thin red mark was a good three and a half feet long and featured little bumps every inch or two. It began to itch, as did several similar but shorter stigmata on the back of my left thigh and on both upper arms.

The funny thing was, I had felt almost nothing underwater—maybe a slight tingle. Nor had I seen anything big enough to be trailing a four-foot-long whip. Riding the current on one of the island's famous "drift dives," I had been careful to avoid bumping into corals; besides, the marks didn't look like coral scrapes.

"These jellyfish here, their tentacles can sting you even when they are cut from their bodies," explained the local teenager who had led the dive. "Most times, you won't feel it until you get out, because warm saltwater numbs it."

I declined his offer to bathe the welts in urine, the favored local remedy.

Applied back at my hotel, unflavored meat tenderizer helped. The papaya enzymes it contains break down the irritating proteins. (The flavored variety includes pepper and other spices that add to the sting.) But the welts were still embarrassingly visible when my live-in boyfriend arrived a few days later.

The dive guide had been right. Named for Medusa, the mythological monster who sported a mane of snakes, the domed creatures popularly known as jellyfish (class: Hydrozoa) trail tentacles studded with cell clusters called cnidocytes.[1] To kill the medusa's prey, typically small fishes and invertebrates, the cluster fires a nematocyst, a coiled thread armed with a barb tipped with a neurotoxin.[2] Cnidocytes remain in residence even after the tentacles they populate are separated from the bell of their medusa. In some species, the cells react to chemical stimuli, even the merest hint of nearby prey. Most, however, respond to touch.[3]

I had likely brushed through a curtain of tentacles detached from a large medusa.

Although Australia's notorious box jelly (*Chironex fleckeri*) packs a toxic wallop that can kill children and even some adults,[4] most jellyfishes along the Mesoamerican Barrier Reef deliver a milder sting. Even with the most toxic varieties, the danger for a diver comes less from the sting itself than from the reckless reaction it can prompt, such as a bends-inducing bolt to the surface in a panicky attempt to avoid the source of the pain.

The medusa trailing the tentacles that stung me may well have been lunch for a snapper or a grouper or possibly a turtle. Lining sea turtles' mouths are small spiny projections called papillae that protect them from jellyfish venom,[5] and jellyfishes are the favorite food of some species. Jellyfishes form a link in the food chain between plankton, which they eat, and marine vertebrates, which eat them.[6] By feeding on jellies, pelagic fish, and reptiles help keep their populations in check; but this habit presents a hazard, because as mentioned in the preceding chapter, to a turtle or a snapper a piece of discarded plastic can look a lot like a jellyfish.[7]

Although aquatic creatures residing on the Mesoamerican Barrier Reef eat jellyfishes, the people who inhabit the land nearby give them a pass. Jellyfish tacos have yet to appear on menus in Cozumel or Roatán. Although mostly water, jellyfishes are fat-free, their flesh packed with omega-3 fatty acids and high in protein in the form of collagen.[8] Some jellyfish species are popular in eastern Asia. From Korea to Taiwan, they appear primarily as a side dish, sliced into ribbons and tossed with chile paste, chopped green onions, and toasted sesame seeds. The flesh has no flavor of its own, but it absorbs strong spices, the tang of which it passes along, accompanied by an agreeable crunch.

The animals that we call jellyfishes are cousins of the hard and soft corals and of sea anemones. All are members of the phylum Cnidaria, a name with a particularly apt root, "cnid," Greek for "nettle."[9] In 1982 researchers at the Smithsonian's Carrie Bow Cay Field Station identified 71 species of jellies around the island and near the mainland and concluded that in sheer numbers, diversity, and voracious feeding habits, medusae

Sea nettle jellyfish, photo by Rowan Wymark.

constituted the most ecologically important species in the area.[10] Sixteen years later, a team of Mexican scientists noted 62 different species on a section of the Mesoamerican Barrier Reef off Quintana Roo.[11]

Carrie Bow Cay scientists also identified 45 species of hydroids on their part of the barrier reef.[12] Closely related to jellyfishes, hydroids are masters of mimicry; some form polyp colonies shaped like feathers; others, like tiny candelabras.[13]

Some marine biologists use "medusa" to refer to any member of a class called Scyphozoa, in popular parlance "true" jellyfishes. Others use the term only to signify the bowl-shaped bell that forms the animal's body. Sporting floating bells and stinging tentacles, a particular order of hydroids, siphonophores, are often mistaken for their cousins, but they have one significant difference: Siphonophores consist of colonies of polyps, each performing a specialized function, whereas every true jellyfish is an individual animal.[14]

All jellyfishes begin life as larvae, which result when the male releases sperm into the water and the female sucks it into her bell to fertilize the eggs waiting within. Once the larvae, called planulae, are big enough to swim, even if rather feebly, they make their way out their mother's mouth and float at the mercy of tides and currents. Those that escape being eaten by fish, invertebrates or their fellow jellies settle within a few days, or even hours, on a hard surface—a rocky sea bottom, a handy pier or the coral reef. Once they do, they have no ability to move around on their own. But they can reproduce, albeit asexually, by throwing off miniature medusae.[15] In some species, any polyp can do this, firing off juvenile jellyfish like so many tiny Frisbees; in others, specialized polyps within colonies take on the task.[16] Some jellyfishes can remain in the polyp stage for decades before letting go and becoming medusae themselves.[17]

Awash in nourishing plankton, the tiny medusae drift and grow, quickly reaching sexual maturity and producing eggs and sperm that, when united, settle down as polyps and begin the cycle anew. This dual method of reproduction gives the species the benefits of both. By generating vast numbers of miniature medusae, a polyp ups the chances of survival for the next generation. But because these juveniles are genetically identical—essentially clones—they share the same inherited vulnerabilities to such threats as disease and changes in water quality. Sexual reproduction may be comparatively hit-or-miss, and medusae of most jellyfish species die shortly thereafter; but the offspring have the benefit of two distinct sets of genes.[18]

One particularly fascinating species of jellyfish, *Turritopsis dohrnii*, can reverse the process by changing from medusae back into polyps.[19] Maria Pia Miglietta, associate professor of marine biology at Texas A&M University at Galveston, has focused much of her research on what are popularly termed "the immortal jellyfish." Although only a few millimeters wide, they are proficient predators that stun their prey in the traditional manner, using tentacles and nematocysts that are deadly for their tiny fellow inhabitants of planktonic tides but too small to penetrate human skin.[20]

Even by jellyfish standards, *T. dohrni* are what scientists call "planktonic." Rather than swimming off on their own, they float around on the tides,[21] vulnerable to whatever nature and humans throw their way. And that makes their ability to live backward such a useful survival trick, for both individuals and for the species. Faced with a lethal threat such as physical injury, lack of food or severe changes in water temperature, *T. dohrni* flips over, forms a cyst on what had been its bell, and reverts to its juvenile state. Attaching itself to a handy hard surface, it begins its life cycle anew, like an aquatic version of the fictitious Benjamin Button. Miglietta and her colleagues suspect that the genetic mechanism for this transformation has something to do with transcription factors: the proteins that tell stem cells which switch to turn on to make themselves into nerve cells, muscle cells, skin cells, and so on.[22]

In 2012 John Gurdon and Shinya Yamanaka received the Nobel Prize in Physiology or Medicine for their work on transcription factors. Using these proteins, Yamanaka prompted skin cells in culture dishes to revert to stem cells. This demonstrated among other things that these "pluripotent stem cells" could serve as effectively in regenerating damaged human organs as stem cells harvested controversially from human embryos.[23] Maria Pia Miglietta thinks that the "immortal" jellyfish can do something similar in nature.[24] Do any other species do this? Could we?

Estimates on the number of species of jellyfishes vary wildly, usually ranging in four figures, and thanks to sophisticated gene sequencing are changing all the time. They all share a similar design. Bowl-shaped in many, four-sided in others, the main body handles the major tasks of reproduction, feeding, and digestion, using the same orifice for ingesting prey (almost all jellyfishes are carnivores) and expelling waste.[25] Typically armed with barbed nematocysts like those that caused my welts on the Cozumel dive, the tentacles extend anywhere from 120 feet in lion's mane jellies (*Cyanea capillata*) to no more than an inch or so in the pudgy, even cute, species known as jellyballs, cannonball jellyfish or cabbage heads (*Stomolophus meleagris*). Because their tentacles are so short and tucked up tight under their bells, cabbage heads are harmless to people. Swimming through a bloom of them is like navigating a sea of floating softballs.

Some common siphonophores are another matter. Take the Portuguese man-of-war (*Physalia physalis*) with its striking iridescent blue float and pain-inducing tentacles.

Although jellyfishes can swim weakly but efficiently by contracting their bells to expel water and deploying the flaps around the edges, they are largely opportunistic feeders. As the medusa drifts through the sea, its tentacles snare prey, which they stun with a neurotoxin. (The chemical composition of the toxin varies from species to species.) Once the wounded animal begins to leak amino acids, the tentacle sends out a stronger, lethal blast of toxin and shortens, bringing the hapless creature to its mouth.[26]

This method of feeding consumes relatively little energy, but otherwise it is not very efficient. Jellyfishes waste enough of their meals to make medusae attractive hosts for parasites. Some jellyfishes harbor entire colonies of tiny shrimps, baby crabs, and juvenile fishes under their umbrellas.[27] Certain small denizens of the reef have evolved specifically to make medusae and other gelatinous creatures their homes.[28] Still others cling to the tops of medusae, hitching a ride along the reef.[29]

It is their gelatinous quality that makes jellyfishes fascinating to observe. Seen on a night scuba dive, they float through a flashlight beam like interstellar aliens. With the light turned off, many flash eerie green bioluminescence. My favorite place to watch them, however, is at large public aquariums, from the other side of a pane of glass. The walls of the tanks are often black or dark blue, the better to show off the wispy creatures within. Artful lighting catches their delicate features. And there's no danger of being stung.

Some jellyfishes are almost completely transparent, like the 14 varieties of moon jellies (genus: *Aurelia*), which appear as disks tinged ever so faintly blue, rimmed in a faint band of white. In the center is a delicate four-lobed pattern—the animal's reproductive tissues.[30] As rare as it is on land, where only some frogs and insects are transparent, beneath the ocean's surface transparency is a popular tool for survival. As coral reef specialist Nancy Knowlton notes: "Many small, tasty animals floating in the plankton wear what amounts to an invisibility cloak."[31]

Being unable to be seen is one thing. But what about being able to see?

In her engaging and informative book *Spineless: The Science of Jellyfish and the Art of Growing a Backbone,* ocean scientist Juli Berwald explained how these enigmatic creatures gather and process information about the world around them. Sitting along the edges of their bells are three or four types of sensory organs.[32] Some of these rhopalia register light and dark; others, water current and gravity.[33] Still other rhopalia contain statocysts, sensors that orient the animal in three dimensions. Moon jellies also possess cilia, tiny hairs that detect the speed and turbulence of the current.[34]

All these sensory cells combine to enable jellyfishes to react to stimuli, but without centralized brains, they have nowhere to process the information.[35]

Yet when it comes to survival, jellyfishes can be far more adaptable than their comparatively brainy neighbors on the reef. Most species of jellyfishes benefit from warmer, more acidic seas, and they tolerate pollution that would do in their coral cousins.[36] Hard coral polyps need to form calcium carbonate exoskeletons, something that gets harder to do as the ocean's pH drops. And although jellyfishes are not fazed by rises in salinity, the turtles that help keep their population in check are. Turtles' saline-secreting glands have to work harder as the ocean becomes saltier, even if the number of jellyfishes in the turtles' diets remain the same. (Estuarian crocodiles, sea snakes, marine iguanas, and many sea birds also possess salt-secreting glands.)[37]

Overfishing, rising water temperatures, increased nutrients from agricultural runoff and coastal development all seem to be tipping the oceans' ecological balance toward jellyfishes.[38] Although jellyfishes don't harm the reef directly, as smothering algae does, they do damage it indirectly. Jellyfishes compete with juvenile fishes, invertebrates, and crustaceans for eggs, larvae, and other food.[39] By eating the juvenile snappers and groupers that would have grown up to eat *them,* and the juvenile parrotfishes that would have grown up to keep the corals free of algae, overabundant jellyfishes can further disrupt the threatened ecosystem.

Both adult jellyfishes and larvae travel on tides and currents and as stowaways in ships' ballast water. By flushing its tanks offshore, even far enough out to be beyond laws prohibiting such discharges, a freighter coming from Japan can introduce a whole new species to the California coast.[40] Under conditions favorable to them, jellyfishes can appear in "blooms" containing millions of individuals. These blooms can stretch so far that scientists need to use aerial drones to map them.[41] To large predators, one of these blooms is an all-you-can-eat buffet. Leatherback sea turtles, for example, consume three-quarters of their weight in jellyfishes daily.[42]

But for humans, jellyfishes can create problems. Consider the 150 million people who are stung worldwide every year.[43] Beaches in Australia frequently have to close during the potentially most profitable summer months due to influxes of deadly box jellies. Beyond delivering stings ranging from merely irritating to lethal, jellyfishes weigh down and often tear the nets of fishers,[44] and their stinging cells can contaminate an otherwise profitable catch.[45] Jellyfishes have been known to invade aquaculture operations, consuming the farm-raised fishes and shrimps.[46] Blooms can clog the desalination plants that produce fresh water for desert countries like Oman and Israel.[47] Jellyfishes also can jam the cooling systems of coastal power plants.[48] An invasion of jellyfishes even shut down the cooling system of a U.S. Navy ship off Australia.[49]

Nonetheless, jellyfishes do contribute to the ocean environment. Some species are so pollution-tolerant that they can thrive around oil spills, where they churn the surface, breaking up the slick into little droplets and helping cleanup efforts further by snagging these in their mucus, which is rich in the nitrogen that oil-eating bacteria need.[50]

In balance, however, to anyone who cares about the health of the Mesoamerican Barrier Reef or corals around the world, jellyfish blooms are a sign of an ecosystem in trouble. By competing with pelagic fish and other organisms for food, and by devouring the larvae and juveniles of these species and of reef-cleansing herbivores like parrotfishes, they contribute to the cycle of decline. If jellyfishes keep increasing, biodiversity will decrease. Some marine scientists think the escalating growth of jellyfish populations is being caused by human activity; others, that it is part of a natural cycle.[51] But even if the attention-grabbing appearance of miles of jellyfishes bobbing on the surface of the seas are part of such perpetual "bloom and bust" phenomena, humans are doing so many things that disrupt the ocean environment that whatever compensatory mechanisms caused natural escalations to subside in the past may not be able to work now.

6

~~~~~~~~~~~

HANGING ON FOR DEAR LIFE

Creatures Anchored to the Reef

In order for hard coral larvae to build colonies, they have to stay put. Encased in their calcium carbonate exoskeletons, the polyps extend their tentacles to snag food, but the polyps never move from their stony tubes. Other creatures, among them the hard corals' cousins the soft corals, are able to bend and sway with the tides and currents. Unlike hard coral polyps, which generate calcium carbonate cups to anchor themselves to their colonies, soft coral polyps produce soft bases to fasten onto theirs. Soft corals attach themselves permanently to the reef but enjoy the benefits of flexibility. To the casual observer, many of them look like plants. Although soft corals lack the roots that anchor plants, these diverse and fascinating animals are welded in place, making them vulnerable to the same hazards as their stony relatives, from hurricanes to pollution to overfishing.

Among the most captivating denizens of the Mesoamerican Barrier Reef are the gorgonians (*Alcyonacea*). Although they derive their name from the Gorgons (Stheno, Euryale, and Medusa), three monstrous sisters of Greek mythology cursed with writhing vipers for hair and faces so horrifying that even catching a passing glance at one turned living creatures to stone (literally petrified),[1] gorgonians reward close inspection. Shaped like fans, ostrich plumes or a collection of long, furry tails, gorgonian colonies can sport intense hues reminiscent of sixties graphic artist Peter Max—acid green, Day-Glo orange, shocking pink, neon violet. In the western Caribbean, they inhabit depths from just below the top of the reef to thousands of feet down.[2] In the Atlantic's colder waters, scientists have found them as deep as 3,000 feet.[3]

For reasons that scientists have yet to figure out, gorgonians are much more common in the Caribbean than they are in the world's other tropical seas. In waters between 15 and 30 feet deep along the cut between Carrie Bow Cay and South Water Caye off Belize, these soft corals are so dense and diverse that this part of the reef has earned the nickname "Gorgonian City."[4] Working at the Smithsonian Institution's field station on Carrie Bow Cay, marine biologist Katherine Muzik identified 13 different genera,[5] the plural of "genus," the taxonomic category between "family" and "species."

Like hard corals, the gorgonians—sea fans (suborder: Holaxonia), sea plumes (*Antillogorgia bipinnata*), sea whips (*Nicella schmitti*), and the like—are colony-forming polyps; but they take a different morphological approach to life on the reef. Rather than the six tentacles wielded by hard coral polyps, gorgonians are armed with eight (thus their alternate name, "octocorals"). Hyper-effective filter feeders, they operate collectively, maximizing both the amount of food they snare and the efficiency with which they use it. As polyps settle onto a hard surface (often living or dead stony coral), rather than encasing themselves in calcium carbonate exoskeletons, they proceed to bud off replicas, meanwhile forming a collective external structure from a tough but flexible substance called gorgonin, similar to human fingernails but with an added punch of bromine and iodine.[6]

Each colony grows by budding. A polyp will produce baby polyps that in turn produce more polyps, all descended from the original colonist. As that original colonist buds off new polyps, the skeleton grows, spreading into a fan shape, a net, the feathery fringes of a sea plume or the waving tubes of a sea whip. Individual polyps also contribute bits of calcium carbonate, called spicules, laid down over the flexible common skeleton. The whole mass is sheathed in slimy living tissue in which the polyps embed themselves.[7] The polyps of any given colony share a simple nervous system, so that touching one part of a sea whip, for example, will cause the surrounding polyps to retract. The colony also shares a digestive system; the nutrition taken in by one polyp is absorbed and circulated to feed the rest of the colony.[8]

While most hard corals feed at night, most soft corals are day feeders.[9] With their tentacles extended, they look fuzzy, in some cases even furry. But snorkelers and divers who succumb to the temptation to pet a sea whip, for example, receive an unpleasant response. Armed with stinging cells (nematocysts) to disable their tiny prey, some gorgonian polyps deliver an odd, itching sensation—not painful, but certainly not soothing.

In addition to budding off baby polyps, gorgonians reproduce sexually.[10] A colony that begins with a male or female polyp retains its gender

Sea fan and sponges, photo by Ben Phillips.

for life. Several times a year, male gorgonians release a blast of sperm, some of which manage to find their way to a female colony of the same species and fertilize the eggs attached to the cells that line the tubes of its polyps. The fertilized eggs quickly become tiny planula larvae. Floating until they reach a suitable hard surface, the tiny percentage that avoid being eaten along the way latch on, become polyps, and begin growing their own colonies.[11]

Soft corals orient their colonies so that wave surges and water currents deliver their planktonic prey directly to them. This is why they are so abundant around gaps in the reef, like South Water Cut off Belize, where the tides, although mild, wash fresh food to them four times a day—twice from the reef and the open sea and twice from the lagoon and mangrove roots.[12] Gorgonians also abound on the edges of walls, the places where the rim of the reef drops sharply, sometimes hundreds of feet. Here regular currents carry plankton and debris past waiting tentacles set perpendicular to the water's flow. A quick and simple way to judge the direction and strength of the current along a wall is to observe the direction and angle of the sea fans that populate it.

Similar to hard corals, gorgonians form cozy symbiotic relationships with algae, which populate the tops of the polyp tubes and provide the colony with sugars through photosynthesis. In return, the algae receive minute scraps of prey and metabolic by-products from the corals. These nutrients help fuel the photosynthesis in a mutually beneficial cycle.[13]

Along the Mesoamerican Barrier Reef, gorgonians are so successful that the main factor limiting their spread seems to be a lack of sturdy surfaces to which larvae can anchor and form colonies.[14] They need a foothold that mud flats and sandy bottoms can't offer. Gorgonians seem particularly fond of shipwrecks, particularly those from the last couple of centuries, where the metal surfaces provide secure footholds. Although gorgonians also benefit from the death of hard corals, because the resulting lifeless mounds provide ideal real estate free of competition,[15] they perform a valuable function in the reef ecosystem by sheltering small fishes and other marine organisms. Some species of oysters, as well as the basket starfish, perch themselves on sea fans while they conduct their own filter feeding.[16] A close inspection of a sea fan can reveal dozens of other aquatic creatures. There's another way that gorgonians contribute to the ecosystem of the reef: When a colony dies or when part of it breaks off, the decay of the skeleton's living sheath releases calcium carbonate spicules that become part of the Caribbean's fine white sand.[17]

Gorgonians have at least one predator, the flamingo tongue snail (*Cyphoma gibbosum*),[18] a stunning-looking creature with a roughly oval shell covered by a rosy gold mantle decorated with irregular black circles. Measuring slightly over an inch, these mollusks attach themselves to gorgonian colonies and use sandpaper-like tongues (radula) to graze on the polyps and other living flesh, leaving patches of bare skeleton. Attacked by only one or two flamingo tongue snails, the soft coral can repair itself, covering the scar with new living tissue and populating this with fresh polyps. But if several snails begin feasting at once, thus clearing a wider swath, opportunistic species may invade the colony and take up home on the exposed skeleton.[19]

One of these opportunistic species is fire coral (genus: *Millepora*). Once it gets hold of a patch of exposed skeleton, it can spread by consuming tissue and polyps, eventually killing the colony.[20] Some fire coral species resemble hard corals, shaping themselves in solid wavy curtains like lettuce-leaf coral or in pointed branches like staghorn. But in fire corals, the surface looks smoother because the polyps are tiny. Although they don't form cups, they pack a painful toxin, which combined with their sharp edges can result in a long and unwelcome souvenir of a tropical vacation.[21]

Another "false coral," black coral (order: Antipatharia), is a more pleasing resident of the Mesoamerican Barrier Reef. Black corals form colonies that resemble small bushes with delicate stems, but their thicker branches take well to polishing.[22] This quality has made black corals so popular in jewelry[23] that nowadays they are rare above depths of 100 feet, although some colonies in the deep ocean rank among the most ancient marine creatures ever discovered. Scientists estimated that one colony was 4,265 years old.[24] (Black corals' relatives the red and pink corals that have been prized in jewelry since ancient Egypt are not native to the Caribbean.)

Sea anemones (order: Actiniaria) rival gorgonians for beauty. Resembling the garden flowers for which they were named, each anemone is an individual animal that functions like a giant coral polyp, only without the calcium carbonate shell of hard corals or the collaborative skeleton of soft corals. Anemones range from a few inches to several feet across.[25] A sticky foot anchors the animal to the reef and supports the fleshy column that forms its body, which may be snugged inside a crevice for protection. Atop this column is a slit-like mouth surrounded by tentacles, often in enchanting shades of blue, green or purple. Depending on the species, the tentacles may be little nubs, half-foot-long whips or something in between.[26] Each of these tentacles is armed with stinging nematocysts to paralyze prey.[27] Although most small fishes and invertebrates are fair game, some species on the Mesoamerican Barrier Reef have developed symbiotic relationships with anemones. Cleaning shrimps not only keep their host anemones free of debris; they post themselves at the tips of tentacles, advertising their parasite-removal services to passing fishes.[28]

Even more remarkable is the relationship between the tricolor anemone (*Calliactis tricolor*) and the starry-eyed hermit crab (*Dardanus venosus*). Approaching an anemone of suitable size, the crab taps it gently, signaling it to relax its hold on its chosen surface. Once it does, the crab picks up the anemone and places it near its shell. Using its tentacles, the anemone climbs aboard, and the crab trundles off, with the anemone figuratively riding shotgun. The crab gets the protection of those poison-tipped tentacles, and the anemone gets ever-changing waters from which to filter food.[29]

Anemones lack brains, but they have nerve cells capable of collaborating to snare food, trigger the mouth to open and prompt the gastric system to digest. In addition to the nematocysts on their tentacles, anemones have curtains of stinging cells inside their bodies to finish off any prey that survives the initial attack.[30]

An anemone's nervous system also tells it when it's time to move on. Although it is almost as rooted to the reef as corals are and prying loose an anemone is difficult, even with a knife, in a pinch the animal can peel itself off and go in search of a better substrate to latch onto. Anemones seem to prefer hard but rough surfaces.[31] Relocating may involve leaving bits of the sticky foot behind. In some species, these fragments can grow into new individuals.[32]

The more common ways for anemones to reproduce are asexually, by budding off little anemones, and sexually, by releasing sperm and eggs that unite to become larvae.[33]

In one genus, *Parazoanthus,* a little anemone resembling a button mushroom, larvae settle on the rims and walls of sponges, where real estate is more available than it is directly on the crowded reef. Some scientists call these tiny anemones unalloyed parasites, but others point out that these aquatic squatters provide protection from fish that like to feed on sponges and can tolerate their relatively mild defenses. *Parazoanthus* may be tiny, but they pack a powerful poison. Because they adopt coloration that contrasts with that of their host, angelfishes and filefishes know to find their food elsewhere.[34]

Sponges (subkingdom: Parazoa; phylum: Porifera) are among the simplest and among the most puzzling organisms in the sea. Each sponge is a single multicelled animal; yet each cell acts independently. Sponges have no organs, no circulatory or digestive system, and nothing remotely resembling a nervous system.[35] They consist of a tough outer skin over a central mass riddled with channels surrounded by choanocytes, specialized cells with sticky rims to trap little bits of debris from the water. This water enters through holes (pinacoctytes) or tubular pore cells (porocytes) in the skin. Tiny whiplike hairs (flagella) move these particles toward the sponge's interior. Other cells take on the job of delivering the food particles throughout the organism and hauling out waste. A set of stiff, needlelike cells called spicules act as a sort of scaffolding, giving some sponges their shape.[36] Depending on the species of sponge, the cleansed water percolates out through vents, called oscules, or via a central cavity, as it does with vase and tube sponges.[37]

How sponges "know" how to take the shape of their particular species and set up housekeeping remains a mystery to marine biologists. But somehow they do. Among the do-it-yourself experiments that marine ecologist Charles K. Biernbaum, professor emeritus at the College of Charleston, posted online was one that demonstrated this ability. The process is simple enough to be conducted in a high school biology lab or even a kitchen. Using scissors, cut a living sponge into small pieces. Place these in a square of tightly woven fabric, dip this packet into a small bowl of seawater (assuming the species is ocean-dwelling), and squeeze. Repeat the pressing until the water is faintly colored. Then set the bowl aside in a cool place. If you have access to a microscope, before throwing away the fragments of sponge, press them in a beaker of seawater until the water becomes dark. Make a "wet mount" by placing a drop or two of this water on one glass slide and covering it with another. Examining the result through the microscope, you should be able to see the cells send out tiny threads, drawing each other together. Even if you don't have a microscope, the experiment will yield visible results: After a couple of days you should notice clumps of cells forming in the bowl you set aside.[38]

Like gorgonians, sponges are filter feeders, but they filter out much smaller particles. Some sponges even specialize in bacteria.[39] Nothing works better than sponges at cleaning the waters that bathe the reef. Particulates are literally their bread and butter, with the size and type of waterborne bits depending on the species of sponge. The largest variety, barrel sponges that grow to five feet tall and more, each process hundreds of thousands of gallons of water a day.[40] That's greater than the volume of an Olympic-size swimming pool!

Because they can't move around, sponges can't go looking for mates. Regardless of the species, reproduction is a simple matter of expelling sperm and counting on the water to unite them with ripening eggs. Timing is important, but many sponges are both male and female, eliminating the need for proximity to another member of the same species. Some species take a live-fast, die-young approach, producing lots of sperm and eggs and, thus, lots of offspring. Individuals that avoid being eaten as larvae long enough to land on one of the rare spots where a parrotfish or other grazer has removed a bit of living coral[41] go on to live from a few years to possibly a century or more. That life span is only "young" when compared to that of their reef neighbors like giant barrel sponges, some of them thousands of years old. Producing fewer larvae, these species focus their energy on individual survival.[42]

Despite, or perhaps because of, their success in complex ecosystems, from tropical reefs to profound depths in arctic waters, sponges are an evolutionary dead end. Because no higher organisms have developed from them, biologists have awarded them their own subkingdom, Parazoa, although it contains just one phylum, Porifera.[43]

Some of the world's approximately 10,000 species of sponges grow on mangrove roots, where they help transmit nitrogen to the plant, at the same time absorbing excess carbon and protecting the rootlets from being nibbled on by tiny crustaceans.[44] But most types of tropical sponges favor the fore reef, settling seaward of the reef crest—on the downward slope, the plunging wall and the "canyons" and ledges of spur-and-groove formations. Among the most arresting sights along the Mesoamerican Barrier Reef are lavender vase sponges, tinted an almost incandescent hue, and brilliant yellow tube sponges poking out from under overhangs and the mouths of coral caves. Vase and barrel sponges shelter juvenile fishes, small shrimps, tiny starfishes, little crabs, and worms. I've seen a cluster of yellowish finger coral 18 inches wide growing inside a big brown vase sponge.

Even sponges lacking hollow centers harbor entire colonies of creatures within their channels. By providing abundant clean water and shelter from predators, sponges serve as both nurseries and havens for many of the other creatures of the reef.[45]

Not all sponges make good neighbors, however. Secreting an acid-like substance,[46] boring sponges (genus: *Cliona*) drill into coral heads, replacing their otherwise durable interiors with weaker sponge tissue, rendering the coral more vulnerable to storm surges and other threats. If a wave topples the coral colony, the boring sponge inside survives, but the reef shrinks.[47]

Flashy, distinctive colors protect the sponges that sport them better than camouflage. Some species are poisonous or at least taste nasty. Recognizing them, hungry fishes either seem to know to avoid them instinctively or to learn after one bite that they're off the menu. That warning can even be passed along to other species. Certain sea slugs ingest bits of sponge poisonous to the slugs' predators; then by only partially digesting the sponge cells, the slugs are able to use their chemical defenses themselves.[48]

Bright coloration can warn off human threats, as well. Species collectively known as fire sponges flash a brilliant orangey red and reward any diver or snorkeler careless enough to brush against them with a burn-like jolt.[49] But fire sponges have also brought forth eight antibiotics,[50] and the chemicals that many other sponge species use for defense can have medicinal properties. One of these is the antiviral drug acyclovir, marketed as Zovirax and used to treat herpes.[51] Biomedical researchers understand that the world's oceans, especially tropical coral reefs with their unmatched diversity, hold tremendous untapped pharmaceutical potential. In acting to protect the Mesoamerican Barrier Reef, we will protect a treasure chest of future treatments and cures.

Sponges also have more mundane uses. Despite the cheap availability of synthetic sponges, nothing matches the feel or function of an old-fashioned tub sponge for a leisurely soak, and car owners seeking a mirror finish swear by natural sponges, insisting that synthetic substitutes leave tiny scratches.[52]

In a surprising exception to the assumption that human beings are the only animals to employ tools, some bottlenose dolphins harvest sponges, which they carry on their beaks and use to flush fishes hiding on the bottom. The canny marine mammals even pass this skill along from one generation to another.[53]

Humans have harvested natural sponges since classical times. In the Mediterranean, divers in ancient Greece would cut the sponge a few inches above its base, assuring that the organism would regenerate, thus developing one of the earliest known forms of aquaculture. In the late nineteenth century, when Florida was still a frontier, a promoter named John K. Cheney recognized the commercial potential of the sponges growing just offshore at Tarpon Springs. By 1905 Cheney and his colleagues had attracted 500 sponge divers from the Greek islands to harvest this natural bounty, which brought in $1 million a year—the equivalent of $29 million in 2020. Tarpon Springs even had a Sponge Exchange.[54]

Taken as a subkingdom, sponges live throughout the world's oceans from shallow lagoons to deep trenches, but it is rare to find a member of a species adapted to one depth living at another that is drastically different. Smithsonian Institution researcher John Macintyre and his colleagues made just such a surprising discovery in the late 1970s. Based at the Smithsonian's Carrie Bow Cay Field Station off the coast of Belize, they found a cave in the reef 19 miles to the north, near an islet called Columbus Caye.[55] The sponges the researchers found inside were more similar to sponges in deep ocean trenches than to those populating the same depths in sunny spots along the reef.[56] The cave, which started about 50 feet below the surface of the sea, dated to the Pleistocene Era. After dropping vertically, it turned almost horizontal.[57] Because this cave was completely dark and its water stagnant, it differed from Belize's famous Blue Hole and even from the caves that threaded from the jungle to the sea.[58] The sponges it housed were able to thrive without sunlight.

But the most remarkable feature of this particular cave was the colony of tube worms that crowded its mouth, where the ceiling was covered so densely by worms that the masses of their tubes formed "pseudostalactites."[59] Only about 1 percent of the tubes that Macintyre and his team examined contained living worms.[60] The cave had no known counterparts anywhere in the world.[61]

Since the work done by Macintyre in the 1970s, marine biology has benefited from enormous advances in gene typing and DNA mapping. Yet much about aquatic worms remains a mystery.

"I wish I knew how many different kinds of worms there were," Anja Schulze admitted during a 2016 phone interview.[62]

A professor of marine biology at Texas A&M University at Galveston, Schulze estimated that the world's oceans contained between 10,000 and 12,000 species of worms, only half of them identified.

"There's incredible diversity in those reef environments," she added.

Settling on coral heads or under shelves in coral caves, many species of worms encase themselves in calcium carbonate tubes, similar to those produced by coral polyps, only longer. Two types that are especially interesting to observe are feather-duster worms (family: Sabellidae) and Christmas tree worms (*Spirobranchus giganteus*). Once ensconced in their coral hosts, they never leave. Instead, they send out pairs of spiral-shaped, often brightly colored, gills about an inch high to harvest plankton and oxygen from the water. When the worm senses a threat, like a diver's extended finger, it snaps back its eye-catching plumes.[63]

Marine worms are distant relatives of earthworms,[64] but their lifestyles are as different as their environments. Some species are specialists, picky about their habitat and thus vulnerable to extinction when storms and pollution threaten their immediate surroundings. Others are generalists, able to survive by recolonizing when local conditions change for the worse.[65]

By hunkering down inside the reef, marine worms avoid becoming snacks for fishes, crabs or snails. Most tube worms never leave their coral homes, except in some cases, timed to lunar cycles, rising to the surface to reproduce. The eggs and sperm of one of these, the Bermuda glow worm (*Odontosyllis enopla*) (which, as its popular name suggests, is not native to the Mesoamerican Barrier Reef), combine with symbiotic bacteria to create a green bioluminescence fueled by oxidation. After this eerily enchanting display, the adult worms die as the newly spawned larvae float off to find their own homes.[66] Other worm species stay put in their tubes when it's time to reproduce but tuck themselves in headfirst, leaving their sexual organs free to snag floating gametes.[67]

One group of marine worms, members of the phylum Sipuncula, insinuate themselves into coral heads, preferring those that are relatively new and still retain their original shape.[68] Although they neither bore nor actively dissolve the coral skeletons, they inhabit existing holes; as the worms grow, they weaken the coral. Fortunately, some hard corals can produce their own cement-like substance to mend the damage.[69]

Some species of bristle worms (class: Polychaeta) move around the reef, using scores of parapodia, which look like little paddles, to swim and crawl. Sporting these appendages, as well as heads displaying eyes, tentacles, and stubby feelers, called palps, these segmented worms resemble caterpillars more than they do earthworms.[70]

In this mobile category, fireworms are lethally well-adapted. Employing venom-spiked needle-sharp bristles to ward off their own potential predators,[71] they eat anything they can catch.[72] Fireworms hunt crustaceans, bottom-dwelling fish, and invertebrates and engulf their prey then dissolve it with powerful digestive enzymes.[73]

The worms that Anja Schulze had studied the most live in piles of dead coral that have broken off the reef. These efficient recyclers thrive on the microbial bounty of their habitat. Rather than making solid tubes, they either fashion tubes from mucus or adapt the coral rubble to protect themselves. Some grow to be as much as a foot long.[74]

The challenge to studying these Sipuncula is that getting a good look at the larger examples involves breaking up the rubble, and that usually breaks up the animal, as well. However, nowadays, observation and gross dissection tell only part of the story. Sophisticated microscopes offer glimpses of a worm's bristles or jaw structure that early twentieth-century researchers never would have seen. Advances in imaging technology allow researchers to study the tiniest larvae and to look beneath a worm's skin to examine its muscle structure. And the ability to analyze DNA has revealed that species that look the same may be genetically distinct. Scientists call this "cryptic diversity."[75]

"Sometimes if you take a second look after studying the DNA, you notice differences that you didn't see before or didn't think were important," Schulze explained. "There is much more diversity than meets the eye."[76]

That is true of all the creatures that make up the fascinating, fragile ecosystem of the Mesoamerican Barrier Reef.

7

THE UNDERWATER KALEIDOSCOPE

From Tiny Blennies to Giant Eagle Rays

"I'm a frustrating person to dive with, because I go slow and I'm always looking into cracks and crevices," admitted Ron Eytan, who has used DNA sequencing to discover three new species of blennies, tiny fishes abundant on the Mesoamerican Barrier Reef. "Someone will say, 'Did you see that eagle ray?' And I'll say, 'No.'"[1]

An assistant professor of marine biology at Texas A&M University at Galveston, Eytan, like his colleague Anja Schulze, had an office barely large enough to accommodate his desk and chair, a couple of bookshelves, and two visitors' chairs shoehorned between his desk and the wall; but he spent most of his time in his lab and on the reef. When I interviewed him in January 2016, he explained that his work had taken him to Mexico's Quintana Roo and to Belize, both to the Smithsonian's Carrie Bow Cay Field Station and to Turneffe Atoll, at 30 miles long and 10 miles wide the largest atoll on the Mesoamerican Barrier Reef.[2]

Eytan truly has to look closely to spot the subjects of his research. Threefin blennies (family: Tripterygiidae) reach only an inch and a half in length; an adult clinid blenny (family: Clinidae) may get to four inches; a combtooth blenny (family: Blenniidae), recognizable by its sloping forehead, four and a half inches.[3] One variety of combtooths, redlip blennies (*Ophioblennius atlanticus*), are velvety black with red fin edges and lips;[4] but most other species are drab gray or brown,[5] and all are adept at hiding in coral niches to escape predators such as hamlets, lizardfishes, and lionfishes.[6] The larvae of the blennies Eytan focuses on, *Acanthemblemaria*, like to take over the holes that tube worms have drilled into dead coral.[7,8]

"These fishes are a fantastic model system to address a host of evolutionary and ecological questions,"[9] he told me. For example, one species of *Acanthemblemaria* that specializes in a particular habitat has been able to persist through several glacial cycles, while another, a "habitat generalist," has not. Yet their life cycles and current geographic distribution are practically identical.[10]

Blennies are exceptional in their genetic diversity. Researchers have identified at least 900 species worldwide, grouping them into six families. As is the case with many inhabitants of the sea, scientists are discovering new species all the time.[11]

All blennies are benthic—they live on or near the bottom. All have elongated bodies, and almost all blennies are small.[12] Some species are widely spread. As a graduate student, Eytan was swimming along a concrete wall in the Bahamas when he spotted a type of blenny previously thought to live only around Carrie Bow Cay in Belize, 900 miles away. The same species may well have existed in the vast expanse between the two locations, but no one had seen them.[13]

Other blennies truly are confined to a small area. "Some species are distinct to one island or one section of the reef," Eytan explained.[14] Discovering this diversity raises the issue of whether to protect a very localized species when there are other types of blennies that look identical to the naked eye.

In the sea as on the land, once a species is extinct, it's gone forever—and so is everything we might learn from it and might not learn from a different species that appears identical. Like all little fishes, blennies perform a valuable function in the ecosystem, eating smaller creatures and being eaten by larger ones. But, as Eytan pointed out, they also have significant value to biologists and ecologists because they offer a window into biodiversity, how and why it develops and how to conserve it.[15]

Eytan and his colleagues are trying to figure out how these different species of blennies evolved. Why might these populations be isolated long enough to account for these genetic differences? For whatever reason, populations hadn't inbred in thousands of years. Diversity is not necessarily adaptive, with natural selection favoring one trait in a given location and a different trait in another. Sometimes it is random, and sometimes isolation alone keeps lineages from breeding with each other.[16] Studying the changes in populations of these tiny coral-dwelling creatures may lead to insights into how to protect the rich diversity of the Mesoamerican Barrier Reef and of oceans everywhere.[17]

When I spoke to him by phone in 2019, Eytan told me that the most recent of the three new blenny species he had discovered since joining the faculty of Texas A&M University at Galveston in 2010 had appeared in a surprising location, just north of Tulum, the spectacular seaside Mayan ruins south of Cancún, on a part of the coast thronging with tourists. Along one particular stretch, however, the waves were so rough that no one was swimming. A graduate student accompanying Eytan pulled him over and pointed out the previously undiscovered little fish in just two feet of water. "The student had to hold me down so I could take a picture," Eytan recalled.[18]

After collecting fishes off the reef, Eytan takes them to Texas A&M's main campus in College Station, to the AgriLife Center, established to provide research support for the School of Agriculture, the "A" in "A&M."[19] There he uses "a very expensive machine" to conduct genomic sequencing and then compares the individual animals' morphology (physical structure). He sometimes finds tiny physical differences that he wouldn't have noticed if he hadn't known that the two blennies he was examining were genetically distinct.[20]

The lifestyles of blennies are also worth a look. Combtooth blennies are herbivores, but most of their kin are omnivores, eating both algae and tiny marine animals[21] like crab larvae.[22] Although blennies are too small to be sought after by humans for food, the aquarium trade prizes them.[23]

Phil Hastings, professor and curator of marine invertebrates at the University of California San Diego's Scripps Institution of Oceanography, discovered two new species of blennies off Florida. This led him to study the overall interrelated functioning of these "very interesting little fishes," what scientists call their "systemics." Eventually he investigated their mating behavior. As Hastings described it, the male tends to guard a territory, such as an empty barnacle, and a female comes in and lays eggs there. After the male fertilizes the eggs, she leaves him to guard them until they hatch into larvae. "He can even attract other females to lay eggs if he has room," Hastings explained.[24]

Males often have distinctive courtship coloration, and they engage in a lot of courtship displays. Underwater photojournalists Ned and Anna DeLoach observed and photographed the elaborate mating rituals of blennies off Indonesia. The males jumped, darted, and turned bright colors, risky behavior for animals that normally hide in the reef to avoid predators. The DeLoaches hypothesized that such antics might be necessary to attract scarce egg-laying females, intimidate rival males and indicate that a particular nest was safe.[25]

When corals die, blennies abandon that section of reef. Yet they also disappear from healthy stretches of coral for reasons still mysterious to researchers.

"Sometimes I'll go to a site, and there's a species that's very common, and I'll go back a few months or years later and see very reduced numbers, and it doesn't always seem to do with the overall health of the reef," Hastings told me. "But on damaged reefs, algae-covered reefs where the coral is dead, they won't be present."[26]

As my mother used to say to console me after a romance broke up, "There are plenty of fish in the sea." She was more correct than she may have realized. Even back in the 1980s, before the development of DNA sequencing, marine biologists estimated that there were 21,565 species of fishes worldwide;[27] several researchers found as many as 150 fish species on patches of coral 10 feet in diameter.[28] William N. Eschmeyer, now emeritus curator of ichthyology at the California Academy of Sciences, has been cataloging fish species for decades. In 2015, he and his colleagues confirmed 33,400.[29] Since then, the census has added about 400 new species a year, and the annual number may be rising—471 in 2017, 417 in 2018.[30] Compared to mammals, with 6,399 species identified and extant in 2018,[31] fishes are impressively varied; compared to insects, which scientists estimate conservatively at two million species (with fewer than half of those identified), not so much.[32]

Even before genetic sequencing, scientists recognized that species variation among fishes was often a matter of seemingly superficial differences, like color.[33] A quick glance at a guide to Caribbean fishes reveals several species of parrotfishes, for example. The midnight parrotfish (*Scarus coelestinus*) is dark blue; the rainbow parrotfish (*Scarus guacamaia*) sports green and turquoise scales and golden fins. But both share the same body shape and the same sharp beak adept at snapping out coral polyps.

In their article on the evolution of reef fishes, marine biologist David R. Bellwood and population biologist Peter C. Wainwright noted: "Many fish families do not need reefs. In both evolutionary and ecological terms, coral reefs represent only one of a range of suitable habitats. . . . In contrast, the presence of fishes appear to have been of critical importance in the evolution of modern coral reefs."[34]

Coral reefs are essential to the specific species of fishes that have evolved in the coastal Caribbean, even though they may not be to some of their piscine relatives. The Mesoamerican Barrier Reef offers fishes both food and shelter. Crustaceans, invertebrates, and fishes all abound in this

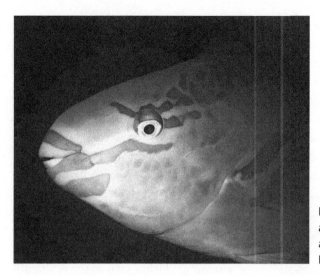

Parrotfish, beak adapted for grazing algae, photo by Ben Phillips.

environment, and holes and crevices in the coral afford fishes places to hide from predators. In exchange, the shrimps, crabs, snails, and worms that share the reef have fish eggs and larvae to sustain them. When the ecosystem is in balance, all the species thrive.[35,36]

Between December 2009 and October 2012, scientists from El Colegio de la Frontera Sur in Chetumal, on Mexico's border with Belize, participated in a genetic survey of fishes along the Mesoamerican Barrier Reef. Recording the DNA "barcodes" of 1,603 eggs, larvae, and adults, they identified 111 families, 231 genera, and 386 species. This topped the numbers logged in a 2005 study. "These results reveal an increase in the diversity of fish in the region of the Mesoamerican Barrier Reef especially for the Mexican Caribbean," reported Lourdes Vásquez Yeomans and Martha Elena Valdez Moreno.[37]

Fishes are vertebrates. Unlike many creatures of the reef but like cows, birds, lizards, and human beings, fishes have internal spines. At one end is a head sporting eyes and other sensory organs, plus a brain to process and react to the information they deliver. All fishes have what's called a "lateral line," a set of parallel tubes that run lengthwise along the sides of their bodies, from their gills to their tails, just below the skin. Pores on the outside allow water to flow through the tubes, which are lined with hairlike projections connected to sensory cells that register pressure.[38] Because they have this handy built-in navigation tool, they can swim close together in schools without running into each other. Other marine animals don't have these.[39] Additional sensors above and below the lateral line pick up the direction and speed of water flow.[40]

Responding to messages passed along their spinal columns, fishes propel themselves through the water by flexing the muscles along their bodies, by undulating their fins or by doing both.[41] Fishes breathe by extracting oxygen from the water via gills, tissues set at the backs of their throats and supported on the outside by thin bony arches.[42] Like other vertebrates, fishes have linear digestive systems, with food taken in at one end and waste excreted at the other.[43]

Apart from these basics, the structure of a particular family of fishes reflects how, what, and when it eats.[44] Some fishes catch and manipulate their food. Among many other carnivorous fishes, barracudas ram their prey with their jaws. Most fishes on the reef make some use of suction in their feeding, many of them exclusively.[45] Fishes such as the brilliantly hued blue chromis (*Chromis cyanea*) that dine on floating zooplankton have upward-facing jaws;[46] those like the boxy spotted trunkfish (*Lactophrys triqueter*) that browse the bottom or the surfaces of coral heads have mouths that point down.[47] If a trunkfish fails to find enough food wandering around, it can aim a strong stream of water to dislodge little creatures hiding in the sand.[48] Wrasses (family: Labridae) have tiny mouths ideal for sucking up the eggs of other aquatic creatures.[49] Possessing mouths that look like lips pursed for kissing, butterflyfishes (family: Chaetodontidae) can pry out crevice-dwelling creatures that flatter-faced predators could never reach.[50] Butterflyfishes and angelfishes (another Chaetodontidae) also prey directly on corals, using teeth as fine as brushes to scrape out the polyps, apparently undeterred by the stinging cells.[51] Trumpetfishes (family: Aulostromidae) hang upside down next to soft corals, their stripes and dots camouflaging them as they wave tranquilly until a hapless little shrimp ventures near their fatal snout and finds itself inhaled.[52] Using their specially adapted fins, batfishes (family: Ogcocephalidae) and frogfishes (family: Antennariidae) crouch on the bottom with tiny wormlike lures protruding from their heads to attract little fishes for their dinners.[53]

Other fishes have small mouths adapted to feeding on plants. Some damselfishes (family: Pomacentridae) even "farm" varieties of turf algae that they find most nutritious, guarding a patch aggressively and driving off invaders and even weeding out forms of seaweed they don't like.[54]

Eels may look like snakes, or in the case of garden eels (subfamily:

Heterocongrinea) like blades of grass rooted in the sand, but they are fishes. The largest variety, green moray eels (*Gymnothorax funebris*), average 6 feet long but can measure 10 feet long and weigh as much as 79 pounds,[55] although they may be only three or four inches in diameter. They coat their brown skins with a yellow-green mucus toxic to predators. A green moray swims by undulating its body. Its dorsal fin is uninterrupted all the way down its back, from just behind its snout to the end of its body.[56] Watching one slither over coral heads is hypnotic. Their super-elongated bodies and streamlined fins allow them to thread themselves inside coral heads and, when they emerge at night, dart quickly to snatch small fishes, crustaceans, and squids. A diver may first notice a chartreuse bulge in the coral, then spot a pointed snout the same color sticking out of the rock four feet below. Repeatedly opening and closing their tooth-studded jaws, moray eels look threatening but are not aggressive, provided they aren't harassed or haven't been taught to expect handouts from divers. Those scary mouth displays are the creature's way of breathing, washing water over its gills as it rests during the day in the coral. Poking a finger into a moray's hole is a very bad idea, however: Those sharp teeth point backward like hooks, and once a moray bites, it can't let go.[57]

Morays seem to display curiosity. One dubbed Waldo has become something of a celebrity on Grand Cayman, extending itself from its eponymous reef to greet divers.

With their underslung jaws and razor-sharp teeth, barracudas (genus: *Sphyraena*) look even more ferocious;[58] and they can lunge aggressively at divers who invade their territories in the middle of the water column above a coral head. But their main danger to humans is to those who eat them: Although their flesh is tasty, barracudas are the most common carriers among 300 species that may harbor ciguatoxin. Produced by single-celled dinoflagellates, members of the same group of algae that cause red tide, this neurotoxin becomes magnified up the food chain, eventually becoming potent enough to cause debilitating symptoms in humans, beginning with nausea and abdominal cramps and progressing to numbness in the mouth and extremities. Some victims of ciguatoxin poisoning, or ciguatera, experience an odd reversal of sensation in which cold things feel hot and hot things cold.[59]

There is no known cure for ciguatera poisoning, and recovery can take weeks. In the early nineteenth century, a British naval physician recommended ingesting copious amounts of "ardent drink." That treatment may have made the patient less miserable but did nothing to address the cause of the distress.[60]

The shape of fishes' bodies reflects how they swim. A bullet-shaped body with a crescent-shaped tail providing most of the propulsion is fast. A stockier body with fins that deliver more of the power trades speed for maneuverability. With their boxy shape, trunkfishes are slow but can maneuver deftly into narrow crevices on the reef.[61]

Possessing a spine gives fishes speed; in general, the more elongated a species, the faster it can swim. But bones also restrict flexibility, making it harder for fishes to tuck themselves into tight crevices than, say, octopuses (see chapter 12). And fishes lack the hard shells that protect crustaceans as they roam around the reef.

To compensate, vertebrate residents of the reef have developed a variety of protective strategies, some downright nasty. Pufferfishes (genus: *Tetraodontidae*) have two defense mechanisms. When relaxed, a pufferfish looks almost comical, a slightly elongated bug-eyed creature with pointy bumps covering its mottled brown, scaleless surface. But when the animal senses a threat, it balloons up to several times its normal size, becoming more than a mouthful even for large aquatic predators.[62] If something big does swallow a pufferfish, it will spit it out immediately; one taste of the nubby skin reveals the presence of tetrodotoxin, a poison that disrupts nerve communication to the brain and is 1,200 times more potent than cyanide, so powerful that even a microscopic amount can kill an adult human being. A typical pufferfish contains enough to dispatch 30 people.[63]

Pufferfishes get tetrodotoxin by consuming bacteria that produce it. Certain newts, toads, and frogs can also harbor tetrodotoxin, but scientists are not yet sure how they come by it. Although tetrodotoxin poisoning in humans is most common in Japan, it has also claimed victims in Brazil and even in Chicago; most species of pufferfishes, including those in the Caribbean, carry this particularly nasty toxin.[64]

The symptoms of tetrodotoxin poisoning could make a hardened assassin blanch. Victims quickly become unable to breathe but remain conscious.[65] In some species of pufferfishes tetrodotoxin is only present during certain times of the year. In most, however, the skin and some internal organs, especially the liver, harbor it year-round.[66]

Pufferfishes are considered a delicacy in Japan. Called fugu, they can only be served in licensed restaurants, where highly paid chefs specialize in separating the poisonous parts from the edible flesh, which they often

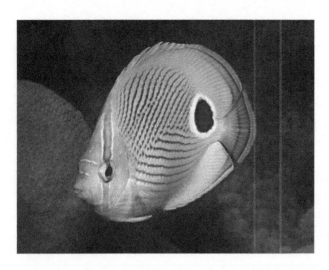

Butterflyfish,
photo by Ben
Phillips.

plate artistically in shapes like birds and flowers.[67] To receive a license to prepare fugu, the chef has to take a special course and pass a final exam that involves eating what he or she makes.[68] Small wonder that restaurants like Fugu Fukuji in Tokyo's fashionable Ginza district charged 9,000 yen—the equivalent of $82.17—for a serving of simple fried or grilled pufferfish in 2019.[69] Fugu can now be raised to be poison-free by controlling what the fishes eat.[70] Still, for some foodies the mere possibility that one's dinner could be deadly adds to the dining experience.

Spotted scorpionfishes (*Scorpaena plumieri*) employ another common piscine protective strategy. Their bumpy gray and tan exterior renders them almost invisible as they huddle amid the coral. But they don't rely solely on camouflage: when it proves insufficient, their sharp, venom-tipped dorsal spines provide a backup.[71]

Groupers (subfamily: Epinephelinae) stay suddenly still when threatened, looking like neutral-hued patches of coral.[72] Many parrotfishes wrap themselves in cocoons of mucus when they sleep.[73] Some species of butterflyfishes sport large black dots on their round flanks; mistaking these circles for eyes, potential predators may conclude that they have encountered a much bigger animal—maybe one that could eat *them*. These false eyes may also fool predators about which direction the butterflyfish might take to escape.[74]

Other species protect themselves collectively. Sergeant majors (*Abudefduf saxatilis*), with their jaunty black-on-yellow stripes, hang out together, sharing their hunting territory with 20 or 30 of their fellows and defending it from interlopers.[75] Plankton-eaters like chromis, goatfishes (family: Mullidae), and silversides (family: Atherinidae) find safety in numbers, with hundreds of seemingly identical individuals flowing beside and above coral heads, moving like sparkling cobalt blue, gold, or silver curtains.[76] Adorned with yellow and blue horizontal stripes, French grunts (*Haemulon flavolineatum*), another group of schooling fishes, are far more attractive than their name, which derives from the noise they make by grinding their teeth.[77] Night-feeders, porkfishes (*Anisotremus virginicus*), a variety of grunts, also protect themselves by banding together as they hunt mollusks.[78] Traveling around the reef in schools makes it less likely that an individual will become a predator's meal. Predators may be as dazzled as divers by the sight of hundreds of small fishes moving in unison.

Brightly hued fishes actually blend in with the brilliant colors of a healthy reef, but they stand out against the drab algae of a reef that's in bad shape and thus are more vulnerable to predators.[79]

Stingrays (suborder: Myliobatoidei) carry their defenses in their tails. These broad, flat fishes nestle on the bottom, fluttering their wide "wings" to flush small crustaceans from the sand. Although otherwise unaggressive, if disturbed by a larger fish or stepped on by an unwary swimmer, a stingray defends itself by whipping up its tail, armed with one or two sharp venomous barbs that inflict painful and lasting wounds.[80]

Southern stingrays (*Hypanus americanus*), one variety that dwells along the Mesoamerican Barrier Reef, are velvety smooth on their white undersides. Their gray tops are more nubby, like coarse linen. A serrated ridge about a half-inch high runs down their backs. With amber eyes set in deep sockets on either side of their fronts, they have no discernible head. When in motion, they ripple through the water. The females are twice as big as the males.[81]

Stingrays' relative, the spotted eagle ray (*Aetobatus narinari*), is one of the most memorable sights on the reef. Gliding above coral heads and through canyons, eagle rays rank among the most graceful fishes in the sea, and their seven- to nine-foot wingspan and even longer tail place them among the largest creatures on the reef.[82] They sometimes rise up out of the water; one jumped onto boat in the Florida Keys, killing a woman on board. The impact was also fatal for the ray.[83]

If eagle rays are amazing, manta rays (genus: *Manta*) are astounding. Depending on the species, their wingspan can measure up to 23 feet.

Twin cephalic fins bracket their mouths, resembling horns and inspiring their alternate popular name, devil fish. Despite their intimidating appearance, mantas are harmless to people, unless they inadvertently hit a diver or snorkeler who gets too close. Feeding on small fishes and plankton, which they scoop into their mouths with their cephalic fins, sometimes while swimming in reverse loops, mantas favor waters where currents concentrate their prey.[84] From May through August, when snapper eggs are abundant, they hang out around Isla Holbox, which lies just 36 miles west of the northern tip of the Mesoamerican Barrier Reef, Isla Contoy.[85]

For a scuba diver the thrill of seeing an enormous ray flap gently by or the hypnotic sight of huge schools of reef fishes makes all the training, all the travel hassles, and all the clumsy lugging of 40 pounds or more of equipment worthwhile. So does the experience of watching the jewel-like individuals darting around the corals. With their indigo heads, magenta gills, and bright yellow bodies, three-inch-long fairy basslets (Gramma loreto) are the showiest fishes on the reef, but they get stiff competition from yellowtail damselfishes (Microspathodon chrysurus), with brilliant blue spots scattered across their matte black bodies. Many fishes change coloration as they mature. For example, juvenile blue tangs (Paracanthurus hepatus) start out bright yellow, and queen parrotfishes (Scarus vetula) go from metallic black to a rainbow of blue, yellow, lavender, and green.[86] Spotted drums (Equetus punctatus) remain black and white but change shape, from a juvenile that seems to be all ribbon-y dorsal and pectoral fins to a more conventional-looking adult, with a recognizable oval body and a shorter but still prominent dorsal fin.[87]

Among the medium-sized residents of the reef, queen triggerfishes (Balistes vetula) rank among the most elegant, with fins sloping gracefully back from their ovoid bodies and two slashes of cobalt blue framing their mouths.[88] Queen angelfishes (Holacanthus ciliaris) display a decorator's dream of blues and yellows, while their cousins, French angelfishes (Pomacanthus paru), sport hip black with yellow and white accents.[89]

Fishes have ears, and some attract or warn their fellows by making noises. By grinding their teeth, grunts make the sound that gave them their name; drums percussively clap the gas bladders they normally use to maintain buoyancy. Several species of fishes, as well as marine mammals, produce sound this way to intimidate rivals or to lure mates.[90] Some fishes and marine mammals even vibrate their gas bladders to sing.[91]

Sound travels farther and four times faster in saltwater than it does in air. When juvenile reef fishes are ready to settle down and stop floating around in the plankton, some are attracted to the coral by the sounds of snapping shrimps.[92]

Fishes also can smell. Marine biologists have concluded that when salmons are ready to mate, scent is what draws them to the freshwater stream where they hatched. On tropical coral reefs, fishes use smell to find food and avoid predators. But pollution from fossil fuels interferes with the senses of smell that many marine creatures rely on to survive, as does the increasing acidity of seawater. Clownfishes (subfamily: Amphiprioninae) placed in water as acidic as predicted by the end of this century lost the ability to tell predators from prey and to find a safe place to settle.[93] Although clownfishes are native to the Indo-Pacific, this demonstration of the effect of ocean acidification on their essential sense of smell may apply to species inhabiting the waters along the Mesoamerican Barrier Reef.

When the sun begins to sink behind the inland mountains, creatures along the Mesoamerican Barrier Reef change shifts. The tiny algae dependent on photosynthesis migrate to the sunny surface at dawn and the darker depths at night, while the little shrimps, crabs, and juvenile fishes that feed on them make a reverse commute, swimming or floating upward through the planktonic cloud as dusk settles, then hiding from their own predators during the day.[94]

Armed with sharp teeth and underslung jaws, jacks (family: Carangidae) hunt in packs, cruising placidly among plankton-eaters by day until the slanting late afternoon sun highlights the school's vulnerable members.[95] Bluehead wrasses (*Thalassoma bifasciatum*), the males sporting azure heads and green bodies separated by two bands of black, retire from a day of munching small crustaceans and seek shelter in the coral, while red-and-white-striped squirrelfishes (family: Holocentridae), their huge round eyes adapted to the dark, swim out from their niches to hunt. With their different feeding schedules, diurnal and nocturnal species can share the same section of reef while avoiding direct competition.[96]

Fishes reproduce sexually, with individuals producing eggs and sperm, which unite either in the open water or within the female's body. In hundreds of species throughout the world's oceans, the male broods the eggs.[97] Among the species common on the Mesoamerican Barrier Reef, seahorse (genus: *Hippocampus*) fathers have kangaroo-like pouches near the bases of their tails for that purpose. Pipefishes (subfamily: Syngnathinae) have something similar along their bellies. The young get protection, nourishment, and a second set of immunities from their fathers.[98]

Male jawfishes (family: Opistognathidae) and cardinalfishes (family: Apogonidae) hold the next generation in their mouths, making these dads resemble squirrels harvesting acorns.[99] Cardinalfish fathers brood their eggs for as long as a week or two, going without food the whole time. This sacrifice is only worthwhile if a large number of offspring result; so rather than deposit a modest clutch of fertilized eggs, which her mate might consider a suitable dinner serving, the female of one cardinalfish species mixes in easier-to-produce yolkless eggs, enhancing the apparent genetic value of the mouthful.[100]

Members of the seabass family (Serranidae), hamlets (genus: *Hypoplectrus*) are hermaphrodites, with the same individual producing both eggs and sperm.[101] Measuring no more than six inches long and notoriously shy, hamlets make up in flashiness what they lack in size and attitude.[102] Some are a dramatic indigo splashed with white; others are a buttery yellow drizzled with turquoise.[103]

In terms of species diversity, gobies (family: Gobiidae) rival blennies. Along with some specialized shrimps and with wrasses, some species, those belonging to the genus *Elacatinus*, perform a service that is fascinating, even startling, to watch. They clean the parasites off large fishes such as groupers.

"Cleaner gobies are striped, and their stripes seem to be advertisements for their services,"[104] Ron Eytan explained.

The big fishes also send visual signals, indicating that they're in the mood for cleaning, rather than eating. For example, black groupers (*Mycteroperca bonaci*) turn white temporarily to signal that it's safe for the smaller fishes to approach and perform this aquatic grooming service.[105]

Witnessing a 35-pound grouper having its teeth cleaned by a tiny shrimp is becoming an infrequent experience, however. As the oceans warm, fishes mature faster, but they don't grow to be as big as they would have if they had been older when they began to reproduce. Therefore, they don't have as many offspring. In 95 percent of the 342 species that ecologists Diego Barneche and Dustin Marshall studied over two years, larger females produced more eggs pound for pound than smaller moms.[106] And the eggs from the smaller females were smaller themselves, meaning that they provided less energy to give the eventual hatchlings a good start in life.[107]

Even on otherwise healthy sections of the reef where small fishes abound, large fishes are rare. Gone are the immense trophy fish photographed on Key West docks in the 1950s.[108] Big food fishes are disappearing from seafood stores. An increasing number of human beings are relying for their protein on a decreasing number of fishes; more and more of these people live at subsistence levels along the coast bordering the Mesoamerican Barrier Reef or make their livings at hotels and restaurants catering to swelling numbers of tourists.

"Overfishing is a growing problem, because people have to eat," Dangriga hotelier Therese Bowman Rath told me.[109]

Although small-time commercial fishers originally resisted the idea of Marine Protected Areas, they now seem to understand that the reserves are essential for the future of the fisheries on which they and their families depend.

"The real competition is the big mechanical operations," Rath explained. "The locals just want to make sure that the rules are enforced for those, too: Don't stop the local fishermen from doing it if you don't stop the others."[110]

And mass collection of food fishes, not all of it mechanized, continues outside the MPAs.

"You don't see a lot of fishing around here," John Tschirky said as we sat in the library at the Smithsonian's field station on Carrie Bow Cay. "Maybe the fish population has been so depleted that the fishing isn't very good. The commercial fishing is mainly based out of a town to the north, Sarteneja. A captain will take six or eight guys out with wooden canoes piled on the boat. When they get to a presumably good fishing site, the guys will disembark, each with a canoe, and fish independently. The captain will receive a percentage of each man's catch. When you see these boats coming back, they'll be really low in the water, so you know they got a lot of fish. I don't know where these guys sleep, the boat's so full."[111]

Whatever the method of catching them, we own the fault in the decline of ecologically important large fishes along the Mesoamerican Barrier Reef: Full-size groupers and snappers, and the parrotfishes that consume the algae that would otherwise smother the coral polyps, are falling victim to overfishing. And overfishing is a response to thoughtless human demand.

8

BIG, BOLD, AND VANISHING

Groupers and Parrotfishes

Groupers (subfamily: Epinephelinea) rank among the largest fishes on the Mesoamerican Barrier Reef. One species, the Atlantic goliath grouper (*Epinephelus itajara*), can weigh in at up to 800 pounds.[1] By contrast, its cousin the Nassau grouper (*Epinephelus striatus*), averages only 50 pounds as a full-grown adult.[2]

To support themselves, Nassau groupers engage in a remarkable example of interspecies cooperation: As mentioned in chapter 1, they hunt in tandem with moray eels. Both favor the same species of fishes. The grouper goes after the prey when it's swimming, the eel when it hides in the reef. When its intended catch vanishes into a coral crevice, a grouper may seek out a nearby moray for help chasing it out again.[3]

I love watching groupers underwater. With their bright eyes and full lips, they seem to make skeptical, if somewhat curious, facial expressions. In one 45-minute dive on Glover's Reef, the atoll 27 miles off the coast just north of Belize's border with Guatemala, I spotted three full-size Nassau groupers cruising solo among the colorful sponges and gorgonians, and one big black grouper (*Mycteroperca bonaci*) hanging just above the drop-off. Named for two eighteenth-century British pirate brothers who used its circle of cays as a base for attacking Spanish ships, Glover's Reef lies just east of the crest of the Mesoamerican Barrier Reef. The atoll displays the healthiest coral ecosystem I had seen in decades. With commercial fishing banned in much of the 86,623-acre marine reserve and sport fishing and diving permitted only in certain areas,[4] Glover's Reef is a model of the ecological benefits of Marine Protected Areas—especially those remote from population centers, even small villages.

My first dive on Glover's was on a site no more than 10 minutes by boat from Isla Marisol, the rustic eco-friendly resort where my dive buddy Terry McNearney and I had booked a cabin for five nights. Live stony corals of many varieties covered about 80 percent of the reef. Anchored to the coral heads were lots of sea fans, large brown barrel sponges, and luminous lavender and white vase sponges harboring tiny shrimps and juvenile fishes; but no particular genus looked like it was crowding out the others. Schools of little blue, yellow, and silver tropical fishes darted among the gorgonians. Bluehead wrasses, purplish blue creole wrasses (*Clepticus parrae*), yellow butterflyfishes (family: Chaetodontidae), vivid green and gold queen angelfishes (another Chaetodontidae), and rainbow and stoplight parrotfishes (*Scarus guacamaia* and *Sparisoma viride*) added to the mosaic of aquatic life. This was what diving at more accessible sites like Cozumel had been like in the early 1980s. As I gazed at the dazzling show, a hawksbill turtle (*Eretmochelys imbricata*) with a shell a meter across swam serenely past my right shoulder.

Two days later, on a beautiful spur-and-groove reef, I saw five hawksbills. Toward the end of another dive, this one at a place called Spanish Bay, an eight-foot nurse shark (*Ginglymostoma cirratum*) lay placidly along the sand in a coral canyon while a remora (family: Echeneidae), slender white fishes that specialize in cleaning sharks and turtles, picked off its parasites.

In the course of eight dives on Glover's Reef, I counted six Nassau groupers and a dozen black groupers, although none of the latter approached the 180 pounds that full-size adults can reach,[5] despite the fact that this was one of the most strictly controlled areas of the Caribbean.

"When I was a child, my mother would have to cut up the fish to fit the pieces in the skillet. Now, you go to a restaurant, and an entire fish fits in one plate," recalled marine biologist Roberto Pott, former Belize in-country coordinator for the Healthy Reefs for Healthy People Initiative (HRI), a multinational nongovernmental organization (NGO) focused on restoring and maintaining the Mesoamerican Barrier Reef. Pott grew up in Belize City. Not only has the size of food fishes declined since his childhood, the populations have, too, as commercial fishermen have sought out the full-grown adults worth the most money at the docks.[6]

"There was this mind-set in some of the coastal communities that there was this big bounty out at sea," Pott told me. "Commercial fishers who had been fishing for 20 years or more were used to seeing thousands of fish, so it wasn't quite accepted when scientists came down from Florida and said, 'If you keep fishing like this, your fisheries are going to collapse.'

The fishers thought 'The Lord will provide. You just have to cast your net on the other side of the boat.'"[7]

But the larger a fish is the more eggs or sperm it produces. That makes bigger, older individuals far more valuable to the preservation of the species, as well as to the customer at the market. As marine biologist Nancy Knowlton noted in her book *Citizens of the Sea,* which summarized the 2010 Census of Marine Life, normally solitary Nassau groupers are especially vulnerable because they return to the same sites every year to mate. Individual groupers sometimes swim as much as 150 miles to congregate with their kind for about 10 days during a full moon between December and March.[8] Most of the year, Nassau groupers wear broad brown and white horizontal bands on their bodies and a brown trident pattern on their foreheads, but they dress up for mating season. Pursued by a swarm of males sporting natty black-over-white mating coloration, female Nassau groupers shoot from 100 feet deep to the surface and release their eggs into clouds of sperm.[9] Within a few days the eggs hatch into larvae.[10]

Local fishers are well aware of the location and schedule of these meet-ups and of the huge potential catches of large fishes they offer.[11] As a result of shortsighted harvesting of the top reproducers, many traditional spawning sites along the Mesoamerican Barrier Reef have become inactive. Groupers pass along the knowledge of the sites from one generation to another, and once all the big fishes have been harvested, there are no more elders to transmit this crucial information. Although there were once 10 spawning sites along the coast of Belize, by 2010 there were only 3. These had drawn 30,000 groupers annually apiece; that number had dropped to only 5,000.[12]

Nassau groupers are listed as "threatened" under the Endangered Species Act, and commercial fishing of them has been banned in U.S. waters, including those around Florida, Puerto Rico, and the U.S. Virgin Islands. Sportfishing of groupers remains legal, but this form of recreational rod-and-hook capture takes a smaller toll.[13] There aren't that many vacation anglers able to plunk down $300 to $1,000 for a day on the water where they and their companions will be lucky to reel in one or two 30-pounders.

Thanks to the efforts of the Reef Environmental Education Foundation (REEF) and the Cayman Islands Department of Environment, one of the Western Caribbean's remaining active spawning spots, Little Cayman's

West End, is now off-limits to fishing and is recovering,[14] but across the Yucatán Strait on the Mesoamerican Barrier Reef, the situation is spotty. Although some spawning sites lie within Marine Protected Areas, elsewhere Nassau groupers can be taken by any method except spearfishing.

Snappers are another popular food fish along the Mesoamerican Barrier Reef. Although the smaller species like the yellowtail (*Ocyurus chrysurus*), the gray snapper (*Lutjanus griseus*), and the aptly named schoolmaster (*Lutjanus apodus*), with its permanently grumpy expression, are holding their own, their larger cousins appear to be declining. Averaging 25 pounds, mutton snapper (*Lutjanus analis*) have become scarce; an angler landing a rare cubera snapper (*Lutjanus cyanopterus*), commonly weighing in at 40 pounds,[15] with a record breaker tipping the scales at 120 off Florida, has a fish story to last a lifetime.[16]

"One of the things we worry about in the ocean is that we've killed off most of the biggest things," Gilbert Rowe, regents professor of marine biology and oceanography at Texas A&M University at Galveston, told me. "We're essentially in the process of wiping out the big animals. We've stopped killing elephants and whales, but what about the big fish? People ought to stop taking these big organisms if only because they're so full of mercury."[17] If we do eat fishes, he added, we should eat the small and medium-size ones.

But eating at least one type of medium-size fishes can be harmful to the reef. Although parrotfishes seldom reach over two feet in length, they make up in eye-catching coloration what they lack in size. In their initial phase, when they can be either male or female, stoplight parrotfishes (*Sparisoma viride*) sport a mosaic of brown and white scales along their flanks above vivid red bellies and tails. Later in life, they all become males and change to vibrant turquoise and lavender with a yellow spot at the base of their tails and a slash of fuchsia marking their gills. Beginning as glossy black, queen parrotfishes (*Scarus vetula*) undergo a similar transformation, but without the dot and slash. Blue parrotfishes (*Scarus coeruleus*) remain true blue, while midnight parrotfishes (*Scarus coelestinus*) are mottled indigo throughout life. With the addition of turquoise, the rainbow parrotfish evolves from a comparatively subdued green and yellow juvenile to a showy mature adult.[18] At night some species of parrotfishes wrap themselves in mucus secreted from an organ on their heads; marine biologists speculate that this DIY sleeping bag protects them from parasites and predators.[19]

If they live long enough, some female parrotfishes undergo a sex change to become what marine biologists call "terminal males."[20] These large individuals are fiercely territorial, driving other males away from their patch of the reef, which they reserve for themselves and their harems of females.[21]

Beyond their value to their own species, parrotfishes perform an essential service to the ecosystem: they keep the bottom clean. Parrotfishes represent 80 percent of the herbivores on the reef, and they spend 90 percent of their waking hours feeding.[22] Using the sharp, slightly over-hung beaks that give them their name, parrotfishes scrape reef rock and coral, making a rasping sound audible underwater. Teeth at the back of the fishes' mouths grind up any ingested calcium carbonate and secrete it as fine, white sand. Parrotfishes' stomachs and guts digest the rest, both algae and polyps.

Algae may look like plants, but they are a different life form; they lack a vascular system and don't have roots, stems, leaves, or flowers. But like plants, algae photosynthesize, turning energy from sunlight into sugars. To thrive, hard corals can't rely on the plankton the polyps can capture on their own; they need species of single-celled symbiotic microalgae, Symbiodiniaceae, to give them a boost of photosynthesized nutrition. But macroalgae smother the polyps, trapping sediments that can build up and slowly bury the coral, depriving them of their planktonic prey and robbing their Symbiodiniaceae of the sunlight essential to photosynthesis.[23]

This enemy of reef-building corals is the favorite food of most varieties of parrotfishes. Although parrotfishes do slurp up living polyps in the process, and some scientists contend that they do as much harm as good,[24] the 2014 report of a 42-year study conducted by the Global Coral Reef Monitoring Network of the International Union for Conservation of Nature (IUCN) concluded that the decline in populations of parrotfishes and other grazers had done more damage than climate change to Caribbean reefs.[25] An ecosystem once dominated by hard corals has become an ecosystem dominated by large algae.

I saw evidence of this firsthand. During the first few of the nine days my dive buddy Terry McNearney and I spent at Hamanasi, a small, five-star eco-resort on the Southern Belize mainland, the weather was so perfect that the director of the Adventure Center recommended that we devote our time to inland tours—Mayan ruins, the jaguar preserve, and other jungle sights that might be inaccessible if it rained in the mountains.

Although February was officially part of the dry season, conditions could be iffy.

Four days later, strong easterly winds prevented the dive boats from leaving shore. The next day, we finally got our chance to see this section of the Mesoamerican Barrier Reef.

Having hired our own guide, we expected to experience the unexpected. And we did. Dropping down to 70 feet, we encountered an underwater dystopia. The visibility was about 50 feet, sub-par by Caribbean standards. Fine silt floated through the water. But what shocked us was the condition of the hard corals. Greenish-brown algae carpeted this whole section of the reef. Sporting what looked like little round leaves on short stems, the algae reminded me of the dollar grass that plagues lawns along the Gulf Coast. A spotted eagle ray sailed eerily past.

Things were better under the ledges of the smothered coral. We saw five spiny lobsters, a sizable Nassau grouper, and both rainbow and stoplight parrotfishes. "Get busy eating!" I told the parrotfishes mentally. Their voracious appetites and sharp little beaks could be the salvation of the reef. The parrotfishes appeared unmoved.

As it grows, a parrotfish becomes more immune to predators. The only creatures a terminal male has to worry about are sharks—and humans. As groupers, snappers, and other food fishes become scarcer, the people who live along the Mesoamerican Barrier Reef may turn to parrotfishes. Although they are not traditionally a part of the region's diet, they are considered a delicacy in Jamaica and some other parts of the Caribbean[26] and are popular in Baja California.[27] Parrotfishes are reportedly mild-tasting, and because they prey on algae and corals, rather than on smaller fishes, they are relatively safe to eat. For example, they are not up-the-food-chain accumulators of poisons like mercury and ciguatera.

In 2015, Guatemala imposed a five-year ban on both the capture and the sale of parrotfishes.[28] The Fisheries Department was one of 350 stakeholders, most of them small-scale commercial fishers, who voted unanimously on the ban. One of its specifications was that no matter where a parrotfish was caught, it couldn't be sold in Guatemala. The populations are being tracked carefully.[29]

"If they see a rise in parrotfish between 2015 and 2020, that will help make the case for extending the ban longer," Ana Giró Petersen, Guatemala in-country coordinator for HRI, explained in 2016.[30]

The case was persuasive. As this book goes to press, parrotfish protection is in place everywhere in the region except for coastal Honduras.[31] The situation was especially complex in the region of Mexico abutting the Mesoamerican Barrier Reef. Compared to Belize, Guatemala, and Honduras, Mexico is huge, with a 2021 population exceeding 130 million[32] and an area of 761,606 square miles—in terms of land mass, the fifteenth largest country in the world.[33] It extends from the Caribbean to the Pacific, and its size, geography and cultural diversity present high hurdles for even the most inclusive and organized coalitions determined to save a species that will, in turn, help save the Mesoamerican Barrier Reef.

It was a challenge, HRI communications consultant Marisol Rueda Flores told me. "Here in the state of Quintana Roo, parrotfish aren't a commercial species. In other words, people don't eat them. But in Baja California, they do."[34]

But the combined efforts of government agencies, local fishing cooperatives, and NGOs like HRI prevailed, demonstrating that education, collaboration, and respect for human communities and the natural environment could protect species that play such a significant role in the ecology of the Mesoamerican Barrier Reef.

9

WHERE HAVE ALL
THE SEA URCHINS GONE?

The Mesoamerican Barrier reef is home to seven species of sea urchins. In the 1980s it suddenly and mysteriously came close to being home to only six.

Although the word "urchin" eventually evolved to denote a homeless or ill-mannered child, sea urchins derived their name from the Old English word for hedgehog.[1] Some short-spined sea urchins do bear a resemblance to the prickly mammals, particularly the West Indian sea egg (*Tripneustes ventricosus*), a reddish brown creature bristling with short white spines.[2]

The hard covering of a sea urchin's body is called the "test." It is often shaped like a dome or a helmet, with a vent at the top for the animal's anus and a larger opening at the bottom for the mouth and the surrounding sticky feet with which the urchin can lumber around the reef and the sea bed.[3]

Unlike hedgehogs, sea urchins lack brains. They are, however, able to sense and respond to changing light. The entire surface of a sea urchin's body acts as a light receptor.[4] If a shadow passes overhead, urchins with spines point some of them toward the possible threat.[5] Each spine is set into a ball-and-socket joint, enabling it to rotate freely.[6]

Sea urchins use their spines purely for defense, not to snare food. And the defense works with almost every potential predator. Queen angelfish have developed a strategy of grabbing a single spine of a long-spined sea urchin (*Diadema antillarum*) and using it to flip the urchin over, judo style, to access the soft spots underneath; frogfish have been known to employ a foot-like fin to step on an urchin spine, then upend the hapless creature.[7]

All species of sea urchins share a distinctive five-sided symmetry, which is obvious in the slightly lobed test of a long-spined urchin and in the five-pointed star or petal design on the flat tops of the two species of keyhole urchin (*Mellita tenuis* and *Mellita quinquiesperforata*) common in the Caribbean. Called sand dollars when they wash up on the beach, keyhole urchins settle into the grassy flats of the lagoon or the chutes between coral ridges, rocking their bodies from side to side to bury themselves in the sand. Although keyhole urchins lack spines, the sand protects them as they filter through it for food. In burrowing urchins, the anus has relocated to the back; in some species, the mouth has moved to the front.[8]

By day long-spined sea urchins tuck themselves into niches and crevices on the reef, leaving nothing visible but a compact cluster of thin black spears. When dusk falls, the urchins march out to browse on their favorite food, algae, sometimes venturing in packs across the sandy lagoon between the reef and the mangroves, ingesting aquatic plants as they go.

Another species, the rock-boring sea urchin (*Echinometra lucunter*), looks similar to the long-spined sea urchin, except that it's much smaller, only one to three inches in diameter, including its pointy black spines. As its name suggests, the rock-boring urchin can do the reef more harm than good. Although it does eat macroalgae, helping keep in check species that can smother the reef, it also settles in by drilling a hole into the coral. When these urchins arrive in number, they can break a coral head apart.[9]

Not surprisingly given the prickly design most sea urchins share, they reproduce by releasing eggs and sperm that unite in the sea to form larvae. Unlike some marine creatures, sea urchins are either male or female, so they can't self-fertilize. Neither do they reproduce asexually, cloning themselves by "budding." They need at least one other urchin of the opposite sex to produce offspring.[10] The success of this process is threatened when the urchins live near shipping lanes, resorts, and fishing villages. Kellie C. Pelikan, a graduate student at Nova Southeastern University in Florida, found that petroleum pollutants degrade the urchins' eggs, reducing fertility and lessening the chances that the larvae that are conceived will survive. It doesn't take a massive oil spill to wreak this reproductive damage; normal fuel leakage from small boat motors is sufficient.[11]

Despite such challenges, the Caribbean's seven species of sea urchins did just fine until well into the twentieth century.

"The sea urchins are among the most conspicuous inhabitants of lagoons and Turtle Grass [*sic*] beds," Hofstra University biologist Eugene H. Kaplan wrote in 1982. "They are abundant and obvious."[12]

The dominant species on the Caribbean's coral reefs at that time was the long-spined sea urchin, *Diadema antillarum*.[13] But then, in 1983 and 1984, one of the most dramatic events ever witnessed by marine science occurred: The long-spined sea urchin all but disappeared from the world's oceans. Coral reefs and seagrass beds where these creatures were once so abundant that their sharp, black, venom-loaded six-inch spines posed a threat to swimmers had become vast graveyards of dimpled light gray lumps circled by the rings of now colorless spines dropped in death. Along some stretches 99.9 percent of the population had died, and in the areas relatively spared, 93.5 percent.[14]

Scientists keep a fairly close watch on the coastal waters of the Caribbean, and fishers, who return regularly to their favorite spots, are quick to notice changes for the worse. One day in mid-January 1983 Harilaos Lessios, a marine biologist working for the Smithsonian Institution's Tropical Research Institute in Panama, was in his office at the institute's headquarters in Balboa, at the Pacific entrance to the Panama Canal, when he received a call from John Cubit, his colleague at the Punta Galeta field station just northeast of the Caribbean entrance to the busy waterway. Cubit announced that he was witnessing unprecedented mortality of *Diadema*. He wanted Lessios to come see for himself.[15]

"Local mortalities of sea urchins happen from time to time," Lessios recalled 36 years later. "Initially I thought this might be another such instance."[16]

But Cubit persuaded him that something unusual was taking place. The two scientists flew out to the San Blas Islands, best known for the elaborate layered textile "pictures," called *molas,* produced by the Indigenous Cunas. Located to the east off Panama's Caribbean coast, the remote archipelago had yet to be affected by whatever was killing the long-spined sea urchins near the canal's mouth and the adjacent coastal towns and industrial areas; so Lessios and Cubit would be able to observe the progress of the epidemic, assuming that it reached that far. They set up quadrants to monitor the pre-mortality populations and watched what happened. Because it took, and still takes, a day flying in puddle jumpers to get to the San Blas field station, the researchers stayed for two weeks at a time, living in cane huts. Beds and a refrigerator were among their scant amenities.[17]

In March, Lessios noticed that the long-spined sea urchins in the quadrants they had set up were dying. They contacted labs and dive shops around the Caribbean and asked them to report anything out of the

Long-spined sea urchin, photo by Alex Poli.

ordinary. Following the path of the currents, the replies came in. Scientists throughout the Caribbean Basin soon noticed that the long-spined sea urchins first began to behave oddly, failing to seek the shelter of crevices in the reef during the day and falling prey to fish that didn't normally eat them. The urchins lost their pigmentation and dropped some of their spines. Within a few days of the onset of symptoms, the long-spined sea urchins were dead.[18]

The epidemic hit the Cayman Islands in June 1983. In mid-July it reached Cancún and Belize almost simultaneously, showing that the pathogen had traveled westward from Panama to the Yucatán Peninsula with the main Caribbean current and then split to travel north and south with the shore currents. It appeared in the Florida Keys in late July and August, the Dry Tortugas in September. By early 1984, labs and dive shops as far north as Bermuda said that they were seeing a rapid die-off of that one species of sea urchin—*Diadema antillarum*—and that all the others seemed healthy.[19] The mass mortality reached Tobago in mid-February 1984.[20]

The long-spined sea urchin had been a feature of the Caribbean ecosystem for hundreds of thousands of years.[21] Now, in an area where the creatures had thrived just a few months earlier, they were gone.

Diadema antillarum are what scientists call a "keystone species," a species whose ecological importance is disproportionately high relative to its abundance and, thus, on which all other species depend either directly or indirectly. The die-off of a keystone species drastically alters the entire ecosystem. In this case, many Caribbean reefs changed from coral-dominated to algae-dominated. Curiously, the other six species of urchins in the same areas were fine.

Lessios and his colleagues quickly ruled out pollutants as the cause. "It is impossible to imagine a pollutant so toxic that it could remain lethal over such a wide area yet so specific that it could affect only one species," Lessios, Cubit, and their colleague D. Ross Robertson reported in *Science*.[22] Furthermore, water temperature and salinity had remained at pre-die-off levels,[23] so a sudden shift in environment could not have been to blame.

Because the die-off began right off the Caribbean entrance to the Panama Canal, scientists suspected that whatever had killed the urchins had traveled by ship before it hitched a ride on the constantly circulating currents. "The fresh water in the canal kills off a lot of marine organisms," explained marine biologist Nancy Knowlton, "but the ships' ballast water is saltwater."[24]

If it came by ship, the pathogen would have been an invader, one against which the long-spined sea urchins had no inherited defenses.[25] Neither would the pathogen have evolved, as most do, to exploit its host but avoid killing it off quickly. Biologically, the long-spined sea urchin wouldn't have stood a chance.

On the Pacific side of Central America, deadly epidemics had struck starfish and other marine species, but they had been localized. Unlike the Caribbean, which is almost a closed basin, the Pacific Ocean is a more open system, so "disease processes in the Caribbean are more catastrophic," Knowlton pointed out.[26]

Whatever killed off the long-spined sea urchins spread so fast in the Caribbean and adjacent waters that scientists were unable to identify the cause. "Nobody was ready for it," Lessios told me, "and every long-spined sea urchin was exposed, so there were no healthy individuals to inject with tissue from the affected ones to see what happened. Also, the state of the study of marine diseases then, and even now, was such that we didn't know as much about the creatures in the sea as we knew about those on land."[27]

Furthermore, identifying the agent that causes a disease is a complicated matter, outside the expertise of most marine biologists and oceanographers. It requires the specialized knowledge of epidemiologists and marine disease researchers, and they weren't around.

Within a year, whatever it was had all but wiped out the long-spined sea urchin all the way from Brazil to the south to Bermuda to the north. Initially scientists thought that populations on the other side of the Atlantic had been spared,[28] but later genomic sequencing revealed that the African urchins that looked identical to *Diadema* were actually a different species.[29]

The impact of the die-off on the Mesoamerican Barrier Reef and other coral communities in the Caribbean was immediate and profound. Macroalgae grew fast. The effects were especially disastrous in areas where populations of algae-grazing fish had been reduced by overfishing.[30] "Full appreciation of the ecological role of this single species of sea urchin developed only after its demise," Harilaos Lessios noted 32 years later.[31]

Combined with other reef stressors, the epidemic indirectly killed lots of reef-building corals. In one section off the Belize coast, three-quarters of the hard corals that had flourished in 1973 became smothered by algae. On the Mesoamerican Barrier Reef and on smaller reefs in the Caribbean, the ecosystems were shifting from coral-based to algal-based.[32] And this was while the reef was confronting plenty of other challenges.

"Algal overgrowth is hardly the only threat that these reefs face," Nancy Knowlton wrote in 2001. "All reef growth is a balance between growth and death, so that any increase in the former increases the future ability of Caribbean reefs to cope with threats that still remain. These threats include other diseases, bleaching associated with global warming, the direct effects of elevated CO_2, and declines in water quality associated with detrimental land use policies, any of which may ultimately prove to be catastrophic."[33]

Along the Mesoamerican Barrier Reef and elsewhere in the long-spined sea urchin's range, overfishing also may have made the population more vulnerable to disease. Removing competitive algae-grazers may have allowed the long-spined sea urchins to settle more densely, which in turn would have enabled the pathogen to spread faster and more efficiently,[34] like typhus in a crowded slum.

One of the remaining puzzles of the epidemic was why it hit only one of the seven species of Caribbean sea urchins. And the lingering tragedy is that it was the species that, next to parrotfishes, was most effective in keeping the reef-building corals free of smothering macroalgae. Long-spined sea urchins are arguably better for the corals because they are less picky and thus eat species of macroalgae shunned by parrotfishes and other grazers,[35] which as researchers James N. Norris and William Fenical put it in the 1980s, at the height of the Cold War, might be engaged in an "evolutionary arms race" as species of algae develop chemical defenses that taste nasty, at least to fishes, and fishes in their turn evolve to develop a tolerance for the chemicals.[36]

Another lingering mystery is why populations of long-spined sea urchins are taking so long to recover. In the areas along the Mesoamerican Barrier Reef in Belize, the population three decades after the epidemic languished at 1.32 percent of what it had been before the die-off,[37] although it appeared to be rebounding at about 12 percent of that dismal percentage a year. On average, the epidemic removed 98.6 percent of the long-spined sea urchins from the environment.[38] Thirty-two years after the die-off, Lessios reported that the most recent surveys indicated that the population had risen to 11.62 percent of what it had been in 1980 but noted that recovery had been sporadic and had varied from location to location.[39] Though depressing, a little more than a tenth is better than the 1/25 that some researchers had thought at first.[40]

Scientists had predicted that this species would rebound quickly because it was so fecund. Long-spined sea urchins spawn throughout the year at every new moon. Females release as many as a million eggs at a time.[41] Presumably because they face less competition for food, surviving long-spined sea urchins are larger than those that were around before the epidemic. Three decades after the die-off, Harilaos Lessios reported that both the mean and the maximum size of the animals had increased.[42] This may help tip the scales in favor of recovery because the larger an urchin is, the more eggs or sperm it produces. However, although their numbers may be greater, those individual sperm are unable to swim any farther than their predecessors. If the eggs released by the nearest female are more than three meters away, they languish unfertilized.[43]

From a perspective of more than 30 years, Harilaos Lessios reviewed the hypotheses scientists had put forth for the slower-than-expected recovery. He concluded the distance between spawning sea urchins might be a factor. But three other explanations struck him as having merit.

First, the pathogen might still be around and continue to kill off long-spined sea urchins. The problem with this hypothesis is that, while there have been subsequent localized outbreaks of the same symptoms, the last one occurred off Florida in 1991.[44]

In a related explanation, the immune systems of this particular species may have been compromised *before* the epidemic, which may be why they were the only species to succumb and also why they have been slow to recover. However, all individuals alive today, except any that hatched from larvae spawned before the epidemic but had not yet settled during it, are descendants of survivors, so they may have genes that convey resistance to this particular pathogen.[45]

Either all the sea urchins around now are descendants of the ones that survived, Lessios told me, or the pathogen died off with its hosts. That's what happened with Ebola,[46] which killed such a high percentage of the people it infected in Africa that it burned itself out.

A second hypothesis that Lessios explored was that the bloom of microalgae in the wake of the epidemic left so little clean coral that *Diadema* larvae had nowhere to settle. This scarcity of available real estate was especially pronounced in areas where other algae-eaters had been overfished. But when he set up four patch reefs and monitored them, Lessios found that what made the difference in recruitment of larvae wasn't the presence or absence of herbivorous fishes, but the presence of another species of sea urchin, *Echinometra viridis*, which cleared space on the reef.[47]

The third explanation he examined was that the long-spined sea urchin's predators have been exerting so much pressure on the now-scarce survivors that a much smaller percentage have a chance to grow to sexual maturity. As evidence, Marine Protected Areas, which have more queen triggerfishes and the like, have fewer *Diadema*. But arguing against this hypothesis is the finding made by scientists who have examined the contents of predators' stomachs and have found no increase in long-spined sea urchins as meals.[48] Lessios suspected that the sluggishness in recovery may be due to a combination of two or more of these factors.[49]

Recovery has been sporadic and has varied from location to location.[50] As recently as 2016, some reefs still had no long-spined sea urchins at all,[51] while others had a remarkable abundance.

Nonetheless, what Nancy Knowlton termed "an undetectably slow recovery" is moving westward across the Caribbean[52] beginning with

Barbados, where pre-epidemic populations of *Diadema antillarum* were the most dense. Slowly, but apparently surely, long-spined sea urchins are repopulating the Mesoamerican Barrier Reef. The Healthy Reefs for Healthy People Initiative's 2020 Report Card noted modest increases in population at sites along the entire length of the reef, with the most encouraging in Mexico.[53] In 2016 I saw some long-spined sea urchins in Belize, a few scattered in the shallows off Carrie Bow Cay, others tucked into crevices on Glover's Reef.

But some scientists are skeptical about the recovery and even the potential of this one species to help save the reef. "I'm not convinced that the species *is* coming back," California State University Northridge biologist Peter Edmunds, who has studied Caribbean coral reefs since 1984, admitted to me in 2019. "That's the bad news."[54]

Edmunds and his colleagues began seeing narrow bands of long-spined sea urchins in water less than 30 feet deep, where the animals were doing a good job of cleaning the reef and creating patches where coral larvae could settle, but the urchins weren't scurrying down deeper to gobble up smothering algae.[55] The population of long-spined sea urchins wasn't dense enough to encourage migration—or reproduction.[56]

"There's a lot more going on in reefs than can be addressed simply by bringing back the sea urchins," Edmunds added. "If you were to bring back the *Diadema* with the snap of your fingers, it wouldn't be enough to turn the reef from an algal-dominated to a coral-dominated ecosystem."[57]

Warming of the oceans is an important factor in what's happened to the world's reefs, and addressing that global phenomenon requires global action. "A really big challenge of coral reef biology now is that when you look at what affects the reefs, it's a combination of local, regional, and global causes," Edmunds told me. "Things are degrading because of these larger-scale phenomena that local events have nothing to do with."[58]

Tragic though the deadly long-spined sea urchin epidemic was, both to the species and to the Mesoamerican Barrier Reef, especially at a time when it was already under threat from human causes, the die-off constituted an unplanned experiment, demonstrating what can happen when an entire species is removed from an ecosystem. One observed result upended assumptions about competition: Despite being given the opportunity, no other species of sea urchins muscled in on the territory. As mentioned previously, *Echinometra viridis* seemed to facilitate the recovery of *Diadema*, clearing algae from corals, thus giving new larvae places to settle.[59] Clear patches are also appearing near the crevices where the long-spined sea urchins tuck themselves in during the day. Even more encouraging for the reef as a whole, hard coral spat (larvae getting ready to light and become polyps) are also showing up on the spots that both species of urchins have cleaned of macroalgae.[60]

Because the epidemic did not occur in isolation, scientists point out that it was only one cause of the algal takeover. The die-off did join with the reduction in parrotfish and other herbivores from overfishing to create a "top down" advantage for macroalgae. At the same time coastal development, expanding agriculture, and unusually strong hurricanes flooded the Caribbean reefs with nutrients, resulting in a "bottom-up" boost.[61] However, uncontrolled development, poor sewage treatment, and bad farming practices had been going on for decades. Given the sudden burst of macroalgae in the years immediately following the long-spined sea urchin crash, the epidemic remained the primary culprit. [62]

Devastating die-offs like the one that hit the long-spined sea urchin probably happened "many times in the past,"[63] Harilaos Lessios explained. But there were no scientists around to observe and document them—or to analyze their impact on the ecosystem. Besides, how could individual coral colonies on both deep and shallow reefs get to be thousands of years old if severe shifts like this one have occurred periodically?

According to Lessios, the *Diadema* epidemic has brought into question many scientific assumptions, some broader than the received wisdom about competition. "The most general thing the epidemic taught us, or at least me, is that we go out and study the marine environment and we think that what we see now is the result of processes that have been ongoing," he told me. "But maybe some animals that are rare were once abundant and were almost wiped out by epidemics like the one that almost rendered the long-spined sea urchin extinct."[64]

10

STARRING ROLES

Starfishes and Their Kin

Among the sights greeting snorkelers as they glide above the seagrass meadows bordering the Mesoamerican Barrier Reef, starfishes rank among the most arresting. Rust red, regal purple or even delicate pink, they stand out against the muted green of aquatic plants. Large orange starfishes roam the seagrass meadows in search of sponges, which they consume to the point that the only sponges still growing on the floor of the lagoon are the ones that taste bad—or, in one case, a tasty sponge that covers itself with another, nasty-tasting sponge species. The tasty sponge pokes snorkels through its odious cover.[1]

Arguing that starfishes aren't fishes, biologists have been trying to get the public to adopt "sea stars" as the animal's popular name, but with limited success.

Like sea urchins, starfishes are echinoderms, members of the phylum of creatures with spiny skins.[2] Of approximately 1,900 starfish species worldwide,[3] a dozen are common in the Caribbean, including *Asterina folium* and *Asterina hartmeyeri,* which are flattish with five broad, stubby arms, like cartoon renditions of their genus.[4] Some starfishes constitute what ecologists call "indicator species," in that the health of their population reflects the health of the ecosystem as a whole. Echinoderm larvae pass through several sensitive stages, so toxicologists use certain species to study water quality. If starfishes along the Mesoamerican Barrier Reef diminish, the whole reef is in trouble.[5] When an epidemic hit starfishes in the tropical Pacific, corals in French Polynesia suffered. So far, nothing similar has struck starfishes in the Caribbean, but it could.[6]

Starfishes keep certain other species in check in a way that no other predator can. Between June 1963 and July 1968, a young ecologist named Bob Paine roamed the tide pools of Makah Bay in Washington State, dislodging and discarding every starfish he could find. Then he recorded

Starfish (aka sea star), photo by Alex Poli.

what happened: Bearded mussels assumed control, throwing the entire ecological balance of the shallow-water community off balance.[7]

That prompted Paine to coin the term "keystone species," which applies to starfishes in the Pacific Northwest.[8] Although starfishes are not similarly essential to tropical marine ecosystems such as the Mesoamerican Barrier Reef, they are nonetheless important to their health.

Starfishes and their fellow echinoderms share a distinctive body design: five segments or multiples of five.[9] Their family tree began half a billion years ago with a creature that attached itself to the ocean floor and used 10 flexible arms to snag passing prey. Somewhere along the way, most descendants turned over, ditched half their arms and settled in mouth down—or in the case of sea cucumbers, mouth forward.[10] Nowadays, the only echinoderm to maintain its original position is the sea lily, which as its name suggests resembles a flower growing on the reef.[11]

Having a mouth facing the sand, mud or rock presents feeding challenges. Once it had consumed the food right below it, the animal would have to move on. And starfishes evolved a distinctive way of doing that: sticky tube feet propelled by a hydraulic system that allows them to amble along, and to attach themselves to prey, while expending very little energy. Here's how it works: Circling the animal's esophagus is a ring canal with five other canals radiating from it. Spaced along the radial canals are podia—those tube feet. Water pressure extends the podia, enabling the starfish to mosey along the bottom or attach to prey. When the pressure subsides, the feet stay put with no effort but with enough force to pry open a clam, for example.[12]

At this point another remarkable ability comes into play for a double whammy at keeping bivalves in check. Some species of starfishes, including the reticulated sea star (*Oreaster reticulatus*) common on Caribbean turtle grass beds, can push their stomachs outside their bodies and insert them into a crack as narrow as the line made by a number two pencil. By prying open a bivalve just this tiny bit, the starfish can slip its digestive system between the two shells and dissolve the creature inside. All that remains for a passing snorkeler to see is a pair of shells, polished clean inside, with only a bit of membrane connecting them.[13]

In addition to employing them to move around or disable prey, starfishes use their podia to taste and smell.[14] How a starfish uses this sensory data remains one of the mysteries of biology, since the animal lacks a centralized brain and must rely on two rings of nerves around its esophagus to process information.[15]

Because they reproduce externally, each starfish has to produce hundreds of thousands of eggs or millions of sperm to have a shot at a genetic legacy. Although males and females look identical, during spawning season they can sense the presence of a member of the opposite sex, triggering the release of gametes.[16]

Starfishes and their echinoderm cousins share an uncanny feature: the ability to regenerate themselves when damaged.[17] An entire new comet sea star can grow from one severed leg; meanwhile the "parent" animal gradually replaces the lost limb.[18] When threatened, sea cucumbers actually "spill their guts," ejecting their internal organs. With the predator either satisfied or confused by the offal, the sea cucumber sets about growing fresh insides.[19]

True starfishes look like a simplified drawing of a star. Some species have proportionately longer legs, others shorter, but all these appendages appear to grow out of their bodies. When starfishes move along the bottom, they lumber along on their hundreds of sticky podia, their arms almost rigid.[20] Brittle stars, on the other hand, look like little pentagons with a single long string fastened to each side, almost as though they had been constructed out of Tinkertoys. Their common name refers to their propensity to break. People attempting to snatch brittle stars find themselves holding one or two detached legs while the crafty creature scampers away on its remaining appendages to hide out and regrow the lost limbs.[21]

The Mesoamerican Barrier Reef is home to at least 63 species of brittle stars (genus: *Ophiothrix*),[22] and scientists continue to discover more.[23] Brittle stars tuck themselves into crevices, set up housekeeping in conch shells, and cling to the walls of vase sponges, sometimes by the dozen.[24]

Compared to starfishes, brittle stars are uncannily agile, moving across the reef and the lagoon floor sinuously enough to earn them their alternate name, "serpent stars." Lacking a suction function, the brittle star's podia are smaller than those of a starfish, and its arms are bristly. On these spines and feet are sensitive light receptors alerting the animal to even slight increases in intensity, so that it can find a dark, secure niche and wrap itself to blend into the background.[25]

With five triangular jaws, each bearing a column of teeth, the brittle star's mouth would be the stuff of nightmares if it weren't so small. This oral armament is efficient at ripping flesh from carrion, as well as seizing small prey and sifting sand for tiny mollusks.[26] Some species, such as the basket star, drape themselves on soft corals and extend their tendrils to strain plankton from the passing current.[27]

Once a brittle star snags or strains its food, it processes it with impressive simplicity. Above those five strong jaws, a single, short esophagus conveys the meal to a stomach, which serves as the complete digestive system, dissolving the food and circulating the nutrients. Any indigestible bits get expelled back through the mouth.[28]

At the base of each arm are one or two slits lined with tiny hairs, which serve to circulate water into and out of the central disk and to release sperm and eggs. Like starfishes, brittle stars reproduce externally, releasing eggs and sperm during a brief spawning season, usually in the spring or summer, and counting on chance to bring them together. In most species, individuals are either male or female, but they look the same. In the one-in-thousands chance that an egg is fertilized, it develops into a larva, which floats around the ocean until it becomes a tiny simulacrum of an adult brittle star and settles on the reef.[29]

If there were a contest for ugliest creatures on the Mesoamerican Barrier Reef, a certain species of sea cucumber, *Holothuria mexicana*, would win hands down. All it takes is one glance at a wrinkly brown donkey dung cucumber to understand how it got its name, and some of its dozen or so Holothuroid relatives fare little better in the looks department. They tend to blend in with the coral rubble that provides their favored habitat. The sea cucumber's active defensive habits also tend to elicit the "yuck" response. Not only do they eviscerate themselves when attacked, as mentioned previously, some species also send out long, sticky pink, red or white tubes from their anuses to ensnare would-be predators. These Cuvierian tubules can render a small crab helpless. In one species, the five-toothed sea cucumber, the tubules are poisonous, rather than sticky.[30]

Unlike other echinoderms, sea cucumbers are laid out lengthwise, with a mouth at one end and an anus at the other.[31] However, their bodies have a distinctive five-part symmetry when viewed as cross sections. Sea cucumbers have little sucker feet, podia, which enable them to move around, albeit sluggishly.[32]

Sea cucumbers breathe through their anuses, sucking in water, then contracting their muscles to force it into the branches of two respiratory "trees" inside their bodies. In contrast to the animal's outer skin, which is too thick to permit the exchange of oxygen, the little sacs at the end of the branches have permeable walls. Although sea cucumbers lack anything we would recognize as circulation, this system delivers oxygen effectively to the cells.[33]

This method of respiration combines with the sea cucumber's spacious interior cavity to attract other creatures in search of homes. Just wait for the next inhalation, and you can swim right in. Small crabs take advantage of this available real estate, but the most commonly observed residents are pearlfishes, translucent creatures with long, thin tails. Sometimes a pearlfish will enter a sea cucumber's anus head first; sometimes it will wiggle in backward. In either case, the resident comes and goes as it pleases, leaving at dusk to feed and coming home before dawn, and may not be too picky about returning to the same sea cucumber. When it decides to eat in, the pearlfish will dine on its host's gonads. (As with other interior parts, a sea cucumber can regenerate its sex organs.)[34]

Biologists have concluded that while this residential arrangement is common, the sea cucumber receives little benefit and often finds it uncomfortable. This may be why five-toothed sea cucumbers have equipped themselves with five bony protuberances called "anal teeth," which can clamp shut around a potential parasite.[35]

Sea cucumbers lack teeth at their other ends. A ring of between 10 and 30 highly sensitive tentacles encircles the animal's mouth, which sucks up algae, detritus, and sand from the bottom like a self-propelled hand vac. Using enzymes to dissolve the food particles, the sea cucumber's intestines make three loops through its body, then expel the indigestible bits as fecal casings, which eventually break into sand. Hofstra University biologist Eugene H. Kaplan cited a study in which sea cucumbers in a single area of less than two square miles processed between 500 and 1,000 tons of sand annually.[36]

As lumpen and lumbering as sea cucumbers are, sea lilies (class: Crinoidea) are the opposite. Gracefully waving sinuous, amply fringed arms as they filter plankton for food or glide along the reef seeking a new location, they look like they belong to an entirely different category of creatures. Yet like sea cucumbers, sea lilies are echinoderms and are closest in appearance and function to the ancestors they share with their cumbersome cousins.[37]

Also called feather stars,[38] sea lilies are the only members of their phylum with mouths that face upward. The mouth sits in a cup of calcium carbonate plates surrounded by arms in multiples of five, from 10 to 40 in the six species common on the Mesoamerican Barrier Reef. Set along these arms, tiny tube feet on the pinnules, the feathery bits, filter plankton from the passing water and convey it to the animal's mouth via minute grooves along each leg.[39]

Hundreds of millions of years ago, the sea lilies' ancestors attached themselves to the ocean bottom via a stalk, which kept them as fixed in place as hard and soft corals are today. The fossil record shows that they clustered together in dense aquatic groves. But over the eons, many of these animals abandoned their stalks, at least as adults. As the oceans cooled and became less rich with plankton, being able to move from place to place made it easier to find food. Mobility also helped them evade predators. Like brittle stars, most sea lilies spend their days tucked into niches in the reef, emerging at dusk to feed.[40]

Remarkably, though, sea lily larvae spend a portion of their early lives recapitulating the lifestyle of their ancient ancestors. They fasten themselves to the reef with a stalk, which they later abandon[41] as they prepare to take their place among the hundreds of other species that make their homes in the reef.

11

HIDING OUT

Creatures at Home inside the Reef

As a culinary delicacy, the appeal of lobster eludes me. Not that I dislike the slightly chewy white flesh of lobster tail; it just doesn't tingle my taste buds. Boiled lobster strikes me primarily as an excuse to indulge in drawn butter.

But watching lobsters underwater is a different kettle of crustaceans. By day, they're a bit difficult to spot, because they tuck themselves into niches under overhangs, with only two spine-adorned antennae waving in the water or tapping on the coral. Unlike American or Maine lobsters (*Homarus americanus*), the species found from Labrador to the Carolinas, the Caribbean spiny lobster (*Panulirus argus*) lacks pincers and the jaunty cap called a rostrum, having instead two natty rostral horns. Both varieties can reach about two feet from their stalk-mounted eyes to their fan-shaped tails,[1] although due to overfishing of individuals that size, the average adult spiny lobster measures 18 inches.[2]

Lobsters and their cousin crustaceans sport carapaces, articulated exoskeletons covering their bodies like a medieval knight's suit of armor. In the case of lobsters, the cephalothorax surrounds the body and supports the 10 legs, while 8 narrower, hinged segments protect the abdomen (the part that we erroneously call the "tail" when it's dished up for dinner).[3] Spiny lobsters along the Mesoamerican Barrier Reef display dapper stripes, either brown-yellow or reddish-orange to blend with their habitat, and white and black spots on their segmented tails. Operating by reflection, rather than refraction like human eyes, lobsters' compound eyes can detect orientation, form, light, and color.[4] Giving the spiny lobster its name, two spine-studded antennae, often longer than its body, sweep back from the front of the cephalothorax. Sprouting between the antennae, two smaller, forked antennules aid in finding and consuming prey.[5]

Spiny lobsters use their antennules mainly for sensory perception, calling on the hundreds of thousands of chemoreceptors on them and, in lesser concentrations, on other surfaces of their bodies and legs. These chemoreceptors pick up on a wide variety of water-soluble molecules, which the lobster's nervous system then sorts, assuring an appropriate response.[6] Lobsters use their chemoreceptors to locate prey, discover a commodious crevice in the reef, find their way home after a night of foraging, identify potential mates, and even avoid diseased buddies.[7] For example, these chemoreceptors enable Caribbean spiny lobsters to sense something "off" about the urine emitted by one of their fellows sick with the virus PaV1, which kills more than half the juveniles it infects. Although normally gregarious, the healthy lobsters shun the sick individual, practicing a form of "social distancing" that helps them avoid contagion.[8]

For defense from predators, spiny lobsters use their much longer antennae, whipping them around like two-fisted swordfighters or rubbing them together to produce a subaquatic din.[9] Threatened lobsters can combine this underwater racket with acrobatic flips. Noisy lobsters tend to survive longer than quieter ones, perhaps because combining stridulation with other defensive tactics, such as antennae brandishing and tail flips, flummoxes predators.[10]

Nurse sharks, groupers, octopuses, and moray eels all prey on lobsters, but along the Mesoamerican Barrier Reef and in many other parts of the Caribbean, the primary predator is human. Pound for pound, lobsters are the most valuable catch in the region.[11]

Although a lobster's carapace offers passive protection, it has two disadvantages. First, it inhibits movement. (Think of that medieval knight, who needed the services of a squire to mount his horse.) Fishes and other vertebrates, including humans, can swim or run freely. For that matter, so can many invertebrates, including octopuses, jellyfishes, eels, and worms. Lobsters can't. But perhaps a greater drawback of the crustaceans' exoskeleton is that the carapace doesn't expand. As lobsters grow, they get too big for their armor, so they grow a new suit under the old one. When the new one is ready, a hormone triggers cracking along the edges, and the animal discards the too-tight carapace. Meanwhile, another hormone prompts the lobster to take on water, puffing it up so that it can burst out of its old shell. Because the newly naked lobster is especially vulnerable, it stays well-hidden until that more commodious carapace hardens.

When the carapace becomes too tight again, the lobster's endocrine system triggers another molting process. An average Caribbean spiny lobster will molt 25 times in its first five to seven years of life and once a year thereafter.[12]

Once the lobster's carapace gets to be from three to three and a half inches long, the animal is probably mature and ready to breed. Although the prime spawning season is March through August in Florida and Mexico,[13] it extends longer closer to the equator.

For mating, lobsters tend to favor shallow water about 75°F.[14] The male deposits a packet of sperm on the ripe eggs under the female's carapace. She then scratches this packet to release the sperm at the same time she releases her eggs from pores at the base of her third pair of walking legs. Because they lack tails, the sperm can't swim, so the timing has to be precise.[15] Successfully fertilized, her clutch of bright orange eggs attaches to her swimmerets, or pleopods, the little paddles under her abdomen that also help her to swim. The mother lobster then retreats to a secure den, staying there for about three weeks, often not even foraging for food.[16] Once tiny dark eye spots begin to appear in the translucent eggs, she seeks out a place where there is strong water movement and flexes her abdominal muscles to send the newly hatched larvae off on their own.[17]

Spiny lobsters measure between one-sixteenth and one-tenth of an inch long when they hatch. Caught by the current, they migrate vertically, dividing their time between the surface at night and deeper waters during the day. Feeding on plankton, they spend five to seven months drifting through the ocean.[18] Sometimes the currents take the larvae hundreds of miles.[19] Undergoing a series of molts, they begin to look more and more like miniature versions of adult lobsters. They finally molt into a transparent stage called a puerulus. These pueruli can swim independently—and very fast. Searching for a safe nursery habitat, they head toward the bottoms inshore of the reef,[20] often settling in mangrove roots or seagrass.

In choosing their juvenile habitat, the pueruli are picky about salinity, seeking out spots with about 35 parts per million of salt. They also rely on chemical cues from red macroalgae and pressure cues from the water column, preferring depths of less than 15 feet.[21] Once they settle in a comfortable shallow environment, the young lobsters grow quickly, reaching an adult carapace length of three inches in two years.[22] Then they migrate en masse to the reef, trudging together for as much as two or three days across the sandy floor of the Caribbean to the coral reef, where they seek out new homes under ledges or in coral crevices. These may be as deep as 270 feet below the surface.[23]

Spiny lobsters dine late. Once the sun sets, they roam the reef and the seagrass meadows in search of small, slow-moving crabs and sea urchins, snails, clams, and mussels. Lobsters' mandibles are strong enough to crack hard mollusk shells.[24]

Eventually, a Caribbean spiny lobster can grow to 15 pounds or more,[25] but this assumes that it survives its youth. Between 96 and 99 percent of juvenile lobsters never make it past their first year. Predators are one reason; but hurricanes also contribute to mortality, as does disease, especially the aforementioned PaVi virus, which is highly infectious and lethal among young spiny lobsters though rare among adults.[26]

On the other hand, a lucky lobster might live 15 or more years, although no one knows for sure, because scientists lack reliable methods for judging a crustacean's age.[27]

Both juvenile and adult spiny lobsters are social creatures. Residing on the reef, they often share communal dens; during the day it's rare to find just one at home. They also migrate in groups. After December's first *norte*, or cold front, when even some parts of the Caribbean can get too chilly near the surface for their taste, spiny lobsters line up, the antennae of one touching the back foot of the individual in front of it, and march together across the sandy bottom, sometimes covering 100 miles or more in search of deeper, warmer water.[28] Like the V-formations adopted by migrating ducks and geese, this subaquatic conga line reduces drag. It also offers protection. If a triggerfish or other predator threatens, the group "circles the wagons," following its leader's cue to form a round cluster with each individual pointing its antennae outward.[29] Once the migrating lobsters find a congenial patch of reef, they disperse to occupy new dens, often with a roommate or two.[30]

Lobsters, crabs, and shrimps are arthropods, related to insects of the land and air.[31] Like insects, crustaceans display thousands of variations on a basic anatomical design—some 50,000 species. Although most crustaceans are aquatic, some, such as woodlice, live on land. The lobsters, crabs, and shrimps that play diverse and vital roles on the Mesoamerican Barrier Reef belong to a class of crustaceans called Malacostraca.[32]

At first glance, crabs look a lot like ticks. Both have eight legs and broad, flat bodies. But, in addition to being adapted to living in and near the water, crabs lack the pointed mouth parts that ticks use to suck blood

Caribbean spiny lobster, photo by Donald C. Behringer.

from mammals and birds. Instead, the front legs of many varieties of crabs have evolved into powerful claws for seizing, crushing, and tearing apart prey and conveying the resulting bits to their waiting maws.[33] Others, including fiddler crabs (family: Ocypodidae) and the Caribbean king crab (*Maguimithrax spinosissimus*), use these adapted appendages for grazing,[34] including grazing on the harmful algae that has covered so much of the stony corals on the Mesoamerican Barrier Reef. The Caribbean king crab is so good at this that the Healthy Reefs for Healthy People Initiative (HRI) joined with fellow NGO Fragments of Hope and the Mexican fisheries agency INAPESCA to test the effect of placing adult Caribbean king crabs on two patch reefs, one off the Yucatán, the other off Belize. In both locations, the macroalgae and harmful turf algae declined.[35]

Although some crabs are terrestrial, their gills adapted to breathing air, most of the crabs along the Mesoamerican Barrier Reef are aquatic, harvesting their oxygen from seawater. Like their lobster cousins, crabs like to hide in crevices along the reef, but they rarely disturb the corals.

The female of the gall crab (superfamily: Cryptochiroidea; family: Cryptochiridae) does cause apparent discomfort. As a larva, she settles on branching hard coral and irritates the tips, prompting the colony to build its exoskeleton around her.[36] (Eventually the resulting bump looks like the gall that a specialized species of wasp makes on a tree.) Meanwhile, the gall crab feeds on coral mucus and on the detritus of algae and fungi that drift into her tiny den.[37] The opening of her personal calcium carbonate cave may become too tight for the crab to exit, but it will also be too tight for a would-be predator to enter,[38] and males, being smaller, can call on her.[39]

One of the most startling sights underwater is what appears to be a sponge ambling across the coral, seagrass or sand flat. Sponges can't move on their own, but they can if they ride on the back of a crab. Several species of Caribbean crabs camouflage themselves by attaching pieces of sponge and algae to their carapaces. The sponge spider crab (*Macrocoeloma trispinosum*) and the aptly named giant decorator crab (*Stenocionops furcata*) engage in this protective subterfuge. The lesser sponge crab (*Dromidia antillensis*) even sports bristles on its back to help anchor its disguise. In securing themselves to their hosts, the sponges benefit by moving from spots where the filter feeding may be poor to those with richer planktonic opportunities.[40]

Tucked back under coral outcrops, crabs use their claws to defend themselves against threats.[41] An attempt to grab one can result in a painful nip.

Like lobsters and shrimps, crabs molt when they outgrow their carapaces. Many crab species take advantage of this necessity to facilitate mating. A few days before she molts, a female will disperse an enticing chemical to attract a nearby male. Treating her gallantly, he will protect her until she sheds her shell. Then he will turn her over and mate with her.[42]

Hermit crabs (superfamily: Paguroidea) are the crustacean exception. They recycle the sturdy shells of snails and other single-shell marine creatures that have either departed or died. As the hermit crab grows, it has to find a bigger home. If it doesn't move quickly, its soft body parts will be exposed to predators, so it really needs this prefab housing.[43]

Most aquatic hermit crabs, as well as many other decapods, can spend periods of time out of the water, provided they keep their gills wet. The six species of hermit crabs common in the Caribbean range in size from the diminutive flat-clawed and smooth-clawed hermit crabs (*Pagurus operculatus* and *Calcinus tibicen*), each half an inch long as adults, to the giant hermit crab (*Petrochirus diogenes*), which grows to more than 12 inches and prefers the roomy queen conch shells that dot the sandy bottoms of lagoons.[44] Whatever their size, hermit crabs share an anatomical distinction: their abdomens curl underneath them, enabling them to wriggle backward into empty shells.[45]

The Saturday of the week I spent at the Smithsonian Institution's Carrie Bow Cay Field Station, the volunteer station manager, Craig Sherwood, prepared a special evening entertainment. He had been rather cagey about his plan all day, but I noticed after dinner that Linette, the cook, had deposited several cups of leftover white rice on the sand a few feet from the kitchen door. As the last glow of twilight faded behind the mountains across the lagoon, the hermit crabs that lived under the building began to gather for the banquet. Clutching bottles of Belikin beer and glasses of lime squash laced with One Barrel rum, we humans followed suit, peering over the porch railing to see what would happen next.

Sherwood appeared with three large snail shells that glowed white in the dim porch lights. He explained that while snorkeling that morning he had scavenged them from the mouth of a moray eel's den, then scrubbed them clean, transforming them, he said, into prime crab real estate. Watch, he instructed us in tones that echoed the showmanship of an 1890 carnival barker, and we would witness the ancient ritual of hermit crab shell exchange.

After milling around in a mosh pit of a couple of hundred individuals, ranging from thumbnail to palm width in size, several hermit crabs ambled over to examine the new shells on offer. A couple of crabs looked to be in urgent need of larger shells. Others appeared to be merely curious, like couples who spend a Saturday afternoon going from one real estate open house to another with no intention of buying. After a while, fights broke out between pairs and even trios of crabs vying for the same shells. Some of the fights got pretty nasty and prolonged, due to the fact that the pugilists were well-matched in size.

Observing some species of crabs underwater can also be diverting. One of the most entertaining crabs to watch is the arrow, or spider, crab (*Stenorhynchus seticornis*). With its head shaped like the tip of an arrow, its yellow-brown body may be as long as two and a half inches; but its eight skinny legs with their bright violet tips are always at least three times longer, like a subaquatic version of a daddy longlegs or a handheld wire scalp massager. On the reef the arrow crab's range is relatively shallow, 10 to 30 feet. Because they do their scavenging at night, scrambling over the coral with a gait at once comical and graceful, arrow crabs are a rare sight on daytime dives.[46]

Culinarily, my attitude about shrimp differs markedly from my indifference to lobster. Bought fresh off the boat, simple boiled shrimp is one of my favorite dishes, but I also enjoy watching them underwater.

A shrimp has a lot of appendages—for walking, 10 articulated legs arrayed along the front of the carapace protecting both the head and the body, another 6 or more along the tail for swimming or carrying eggs, and 3 pairs of short ones near the mouth for feeding. Plus, a shrimp has two forward-pointing antennae, which may be forked. In some species, their walking legs may be equipped with pinchers.[47]

All this makes for entertaining action as shrimps scavenge among the coral for decayed algae, bits of carrion or living prey such as small worms and crabs.[48]

Snapping shrimps (family: Alpheidae) are often heard before they are seen. Using a specially adapted claw, they make a popping sound so sharp that they are also known as pistol shrimps. After opening and locking this oversized finger, the shrimp clenches its palm, then releases the tension producing a noise that startles predators and stuns small fishes that then become easy prey.[49]

Most snapping shrimps are social animals. They live in colonies inside some species of sponges, each colony with an egg-producing queen and an army of nonreproductive adults charged with defense. When an intruder threatens, the shrimps snap in unison making a din that drives most potential predators away. Those that don't get the auditory hint are dismembered by the mass of claw-wielding crustaceans.[50]

Other varieties of snapping shrimps enjoy a symbiotic relationship with gobies. The shrimp excavates a burrow, which it shares with the fish. When danger approaches, the goby warns the shrimp, and both dart into the hole. The two stay in constant contact, with one of the shrimp's antennae always touching its goby partner.[51]

Divers willing to move slowly and pay attention to the smaller residents of the reef may be treated to the arresting example of subaquatic symbiosis, mentioned in chapter 8: cleaner shrimps removing parasites from the mouths and skin of larger creatures, some of them predators. Of 100 crustaceans that offer tonsorial services on the world's coral reefs,[52] the spotted cleaner shrimp (*Periclimenes yucatanicus*) and the banded coral shrimp (*Stenopus hispidus*), which sports red and white stripes like a barber pole, are among the most common and colorful in the Caribbean. Both use their showy coloration to advertise their services to passing fish. Some cleaner shrimps wave their antennae to indicate that they're available; others assemble at cleaning stations and engage in rocking dances, like the employees in the 1976 movie *Car Wash*. Fishes that want pesky bloodsuckers removed stop by for grooming and indicate by their body language that they are there for peaceful purposes.[53] Even a diver, if calm and patient, can receive a manicure at one of these underwater spas; cleaner shrimps may include dry skin and cuticles in their diets.[54] Some crafty species masquerade as cleaners but cheat their clients by eating their mucus, rather than their parasites.[55]

Like lobsters and crabs, snapping shrimps, cleaner shrimps and most other shrimps on the Mesoamerican Barrier Reef belong to the order Decapoda (10-footed). Mantis shrimps, on the other hand, occupy their own separate order, Stomatopoda, as well as the suborder Unipeltata. Although they resemble true shrimps superficially, closer inspection reveals remarkable differences. First, a mantis shrimp's carapace begins at the rear of its head and covers only the first three segments of its thorax, allowing the animal to rear back and assume a posture similar to that of a praying mantis, the insect that inspired its common name.[56] The mantis shrimp is so flexible that stranded out of the water, it can execute reverse somersaults across the beach to bring itself back to the sea.[57]

Mantis shrimps are among the animal world's fastest and most formidable predators. Equipped with bright red clubbed claws that can function like hammers to break open the shell of a snail or other prey, mantis shrimps can strike with remarkable speed and force. Scientists have measured the strike of the peacock mantis shrimp (*Odontodactylus scyllarus*), native to the Indo-Pacific, as equivalent to a .22-caliber bullet.[58] Here's how the mechanics work: Set just above that clubbed claw, a saddle-shaped portion of the arm stores energy when the animal contracts its muscle, then releases it on command like an archer's bow.[59] Some mantis shrimps have teeth on their second pair of legs, which they use to seize prey, sometimes cutting the victim in half. Even though the Caribbean rock mantis shrimp (*Neogonodactylus wennerae*) seldom grows longer than three inches, it sports a finger with a razor-sharp central ridge that can split the thumb of a diver or snorkeler who disturbs its coral den.[60]

Mounted on stalks providing 360° vision, mantis shrimps' remarkably complex eyes can sense 10 different wavelengths between red and blue, while we humans perceive only 3. Mantis shrimp eyes can also pick up ultraviolet, which we can't see.[61]

We should be glad that these amazing creatures are relatively small, averaging only three to four inches long—and that they confine themselves to the sea. But some mantis shrimp do grow larger, including one estimated to be about 18 inches caught off a dock in the Indian River near Fort Pierce, Florida.[62] Photographs in the *Brevard Times* made it look like the stuff of nightmares.

12

ARMED FOR SURVIVAL

Octopuses and Squids

The first time I saw an octopus in the wild was on a night dive off a pier in Cozumel, Mexico. I was staying at Casa del Mar, a congenial medium-size hotel on the west side of the island a few miles south of San Miguel, the main town. Across the two-lane coast road was the hotel's dive shop, with lockers for guests to stash their gear. Eager to get underwater, my dive buddy and I gave the excuse of checking out our equipment shortly after we arrived that late afternoon, rather than waiting for the guided dive the next day. As the sky darkened over the Cozumel Channel, we suited up and did back-roll entries off the platform attached to the dock.

Thanks to the streetlights along the coast road, the lamps around our hotel's pool and bar area, and the twinkling hulls of two cruise ships docked half a mile away, we hardly needed our waterproof flashlights. We watched a school of tiny silvery fishes dart among the pilings, searching for someplace safe to spend the night. Then we spotted movement halfway up one of the pilings. Something glowing faintly turquoise—no, lavender; no, turquoise—was flowing along it, first up and then down. It was small, only about 18 inches from the tip of one of its eight appendages to the tip of another. An octopus. It moved like no creature I had seen before on land or in the air.

Like their cousins squids, octopuses (not "octopi," the English false plural resulting from the mistaken assumption that "octopus" is Latin, when in fact it is Greek) are invertebrates and lack a shell, which is why they can seem to flow, even when they are grabbing onto hard corals, rocks, and pier pilings with their sucker-augmented arms. When they need to move faster—for example, to escape predators—octopuses suck water into their mantles, then expel it through a funnel that they can point, so that they jet around, arms trailing. Squids use this method of locomotion most of the time, but they also have fins along their sides that can ripple to enable the animals to make complex movements—left, right, up, down, forward, and backward. Set close to mouths ringed with eight arms and two longer tentacles, squids' large round eyes help them snag prey. Unlike octopuses, which are loners, squids in the Caribbean travel in schools, called "shoals," typically consisting of 4 to 30 individuals.[1]

Octopuses, squids, and cuttlefishes are part of a class of marine animals called cephalopods (Cephalopoda), represented by more than 800 species worldwide, including the giant squid (*Architeuthis dux*) that dwells in deep oceans and can reach more than 40 feet from the fins behind its mammoth head to the tip of its longest tentacle. An even bigger relative, the colossal squid (*Mesonychoteuthis hamiltoni*), inhabits the depths near Antarctica. By comparison, Caribbean reef squid, the subspecies most common on the Mesoamerican Barrier Reef, are pikers, no more than eight inches long. Their scientific name, *Sepioteuthis sepioidea,* refers to their resemblance to cuttlefishes (order: Sepiida), which have unique oval interior shells made familiar by their popularity as beak sharpeners in birdcages. For reasons that scientists have yet to figure out, cuttlefishes limit their range to the waters around Europe, Asia, Australia, and Africa; so they don't populate Caribbean reefs.[2]

Neither does another cephalopod, the chambered nautilus (*Nautilus pompilius*), heralded by nineteenth-century American poet Oliver Wendell Holmes Sr. as "the ship of pearl" and "child of the wandering sea." Restricting its wandering to the Western Pacific, it sports an external, chambered shell along with a bristle of up to 90 tentacles, but it is not nearly as highly evolved as octopuses, squids, and cuttlefishes. It hasn't changed substantially in half a billion years, making it a living fossil.[3]

Cephalopods are the literal bluebloods of the animal world. Rather than employing iron to circulate oxygen through their bodies, they utilize a copper-based protein, hemocyanin, so their blood turns blue when exposed to air. Because hemocyanin is less efficient than hemoglobin, the red blood cells that we and most other animals use to circulate oxygen, cephalopods need oxygen-rich water to survive and have never made the evolutionary transition to fresh water.[4] Cephalopods do best in seawater, which has around 32 to 38 parts per thousand of salt with a pH around 8.4, well to the alkaline side of a scale of 1 to 10.[5]

For humans, cephalopods are interesting for two main reasons. The first and longest-standing is as food. Unlike fishes, these marine creatures neither metabolize nor store fat; and because they lack skeletons, they need much more muscle than fishes of similar size. That makes them an excellent source of protein. Granted, being so muscular renders cephalopods tough, but proper cooking turns them into a tasty, and relatively tender, dish. Cuisines from Spain to Thailand feature cephalopodic delicacies, and trendy chefs have gone well beyond the familiar calamari appetizer—breaded and fried rings of squid—ubiquitous at American seafood houses and Italian restaurants.

Cephalopods' other interest to people is as research subjects, especially for studies into the working of nerves. Although they seem at first glance to be almost as unlike *Homo sapiens* as an animal can get, cephalopods have yielded significant insights into how we sense the world around us and how we process and use that information—and even into the nature of consciousness itself.[6]

The scientists who study them refer to octopuses as "octopods" because they have eight arms, and to squids and cuttlefishes as "decapods" because in addition to eight arms, they have two longer (in some cases much longer) tentacles. All cephalopods are mollusks, members of the same phylum as snails and slugs; however, in terms of nervous systems and behavior, they differ dramatically from these distant relatives.[7]

For one thing, cephalopods possess statocysts—a system to sense gravity and a system to sense angular acceleration. If you move in a three-dimensional space, you have to have a way to tell you where you are in all three. In the case of octopuses, one set of receptors helps them perambulate along the bottom; another enables them to jet around the reef. Animals that move in only two dimensions don't need anything like such a complex arrangement.[8]

Octopuses and squids are highly successful predators. They have to be. Although they start as hatchlings, in most species only one twenty-fifth of an inch long, adult squids measure anywhere from half an inch for the smallest, the southern pygmy squid (*Idiosepius notoides*) to 46 feet for the largest, the colossal squid. Octopuses have a similarly dramatic growth curve. In their first three weeks of life, baby common octopuses (*Octopus vulgaris*) double their size. To support increases on this scale, cephalopods hunt and eat everything from plankton to bony fishes larger than themselves.[9] Squids and cuttlefishes regularly feed on fishes.[10]

Cephalopod hunting techniques vary from species to species, but they include such sophisticated strategies as stealthily stalking their prey and even disguising themselves as a clump of algae or a submerged coconut.[11] Once they reach adulthood, cephalopods remain voracious, because their highly mobile predatory lifestyle demands a lot of energy. Unlike us and our fellow vertebrates, they don't store calories from the food they consume. Their activity and metabolism combine in a self-perpetuating loop: cephalopods have to hunt and eat constantly because they move around a lot, and they move around a lot because they have to keep hunting and eating.

Like other predators, squids play a vital role in the ecosystem of the Mesoamerican Barrier Reef by keeping other species in check. Reef squids ingest between 30 and 60 percent of their body weight daily, mostly in the form of shrimps, marine worms, and small crabs and fishes. This means that these squids have to be crafty and efficient hunters—and they are. Their chromatophores, complex organs incorporating five different kinds of cells, allow them to imitate the colors and patterns of seaweeds and soft corals and lurk there while awaiting prey. Although their normal color is greenish-brown, squids can switch to a lighter color, taking on an almost transparent hue of sunlit water when they sense a larger animal, including a snorkeler or diver, nearby. They can even decorate themselves with multiple eye spots, the better to baffle predators or prey. With their two relatively long tentacles, squids grasp the unsuspecting fish or crustacean, then use their arms, which are shorter, to deliver it to their mouths, where they tear it to pieces with their sharp beaks. A tooth-studded tongue called a radula helps convert the resulting morsels into a slurry, which they convey to their digestive tracts.[12]

Caribbean squids spend their yearlong lives in and near reefs. During a literally flashy mating ritual in which a shoal of sexually mature squids swim together in a circle, a male lures the female of his choice from the group by using his chromatophores to pulse an alluring pattern on one side of his body while simultaneously intimidating rival males by displaying a bellicose pattern on the other. (Pattern seems to be key here, since most researchers have concluded that cephalopods are colorblind.) If the visual signals that his intended flashes back indicate reluctance, he strokes her gently with his tentacles until she allows him to come close enough to deposit a packet of sperm inside her body, using an arm that is shorter than his other seven and thus adapted for this purpose. Employing her own arms, the female then deposits her fertilized eggs—as many as 70,000—in a clutch of tough sacs under a ledge. This ritual is literally her last dance; she dies almost immediately. Her partner initially protects the precious genetic deposit from predators but then leaves to hook up with other females before he expires, as well, shortly thereafter.[13]

Depending on the water temperature, the eggs may take as long as eight weeks to incubate. When they hatch, the baby squids head for the shallows, dodging the jaws of hungry little fishes along the way. Those that survive settle into water a meter or less deep, living on plankton until they grow big enough to relocate to patches of turtle grass, where they hide out in water about two meters deep, hovering far enough below the surface to evade birds but far enough from the bottom to avoid medium-size fishes like snappers. Once grown, the adult squids can zoom and flutter as deep as 300 feet before ascending to sections of the reef between 4 and 24 feet to mate and die.[14]

Overall, cephalopods possess the most complex nervous systems of any invertebrates—more than most fishes and even amphibians. To compete with fishes, squids and octopuses evolved elaborate sense organs. Cephalopods' highly developed eyes are their main asset, especially for octopuses. Octopuses' optic lobes are huge, up to seven times the size of their brains. Depending on the type of octopus, each optic lobe contains between 60 million and 90 million nerve cells. That's why these animals are such nimble visual learners, differentiating shapes, seeing how to get into a clear box containing a tempting shrimp and even stacking plastic pool toys.[15]

Although cephalopods are colorblind, they can detect polarized light, including that reflected by fish scales.[16] They can also sense dim light filtering down hundreds of feet from the surface. Their eyes operate separately, so that one eye of an octopus can focus on an aquarium attendant offering a fish and the other on a sea anemone at the other end of their shared enclosure.[17]

Squids and octopuses also use their eyes to help them maintain equilibrium—something that humans do, too.[18] That's why if we feel queasy on a rolling boat, focusing on the stable horizon often helps.

Although like octopuses, squids are primarily visual predators, they also have finely honed motion-detecting receptors arrayed in lateral lines along the sides of their heads. Different from the tubelike lateral lines possessed by bony fishes, squids' consist of rows of tiny hairs that allow them to sense slight water movements, even in murky conditions and at an impressive distance, such as when a potential tasty morsel swims by 10 yards away. The receptors also enable squids to school, moving in a synchronized group without running into each other.[19]

Cephalopods' brains are large, but their brain-to-body mass ratio is modest compared to our own. The average human brain contains between 85 billion and 120 billion neurons.[20] For a 150-pound person, 100 billion works out to a ratio of about 667 million brain cells per pound. By contrast, the brain of an adult common octopus contains about 40 million neurons. For a typical animal weighing 1.5 pounds, the ratio would be about 27 million to one.[21] But the size of its actual brain tells only part of the story.

Two-thirds of an octopus's neurons are relegated to its extremities. An adult common octopus has approximately 520 million total nerve cells, a thousand times what it had as a hatchling. A whopping 130 million go to its optic lobes. Added to the 40 million in its brain, that makes 170 million neurons.[22]

The remaining 350 million are distributed to the suckers along its arms.

Each sucker on an octopus's eight arms has 10,000 chemoreceptor (chemical-sensing) cells—a total of 16 million per animal—and each of those suckers functions independently from the animal's main brain, relying on "detector proteins" to differentiate between something tasty, such as a crab, and something repellent, such as Velcro, which scientists have found that octopuses abhor.[23]

Octopuses have no idea where their arms are relative to their bodies. At the center of every sucker a "satellite" brain analyzes the chemical information the sucker samples and coordinates with the other neural processors in the suckers of that arm to take action. That's why, unlike squids, octopuses can hunt prey that they can't see.[24]

Probing under a coral shelf in the reef, the suckers may catch a tantalizing subaquatic whiff of a small crab hiding in a crevice, for example. Rather than send this information to the brain, the suckers that sense it take action on their own, triggering the octopus's arm to reach under the ledge, snatch the crab, and bring the toothsome crustacean to its mouth for centralized action—injecting a neurotoxin to subdue the crab, then tearing it apart with a sharp, parrot-like beak and consuming it. Caribbean reef octopuses hunt at night, often spreading the webs between their arms to contain a hapless crustacean or fish in a deadly tent while they devour it.[25]

To understand why octopuses need such a large and decentralized nervous system, sit with your elbow on a table and a graspable object—say, a salt shaker—within reach. Keeping your elbow where it is, lower your arm and grab the object, then raise it a few inches. As you do so, think about the movements mechanically; your forearm is the lever, your elbow is the fulcrum, and your wrist and fingers function similarly on a smaller scale. Every element of the action of grabbing the salt shaker involves your bones and joints—your skeleton. And because humans, like other vertebrates, possess that infrastructure, we can direct it all from one central location—our brains—and take advantage of the mechanical efficiency that our bones and joints provide. Now think about trying to grasp and lift an object with a muscle that functions without a skeletal assist—for example, your tongue. Even picking up a shelled peanut with your tongue would demand enough concentration and coordination to impress fellow patrons at a bar.

To make it to adulthood, an octopus has to be accomplished at hiding, as well as hunting. As a method for baffling predators, cephalopods use "ink," which they carry in sacs near their anuses. Varying in color from black for octopuses to blue-black for squids to brown for cuttlefishes, which lent a version of their scientific name to the particular shade "sepia," the ink is almost pure melanin, the same substance that gives color to our own skin and hair. When alarmed, these canny mollusks mix the ink with mucus and expel it as a slimy smokescreen. Some even create an inky likeness of themselves, tricking a hungry fish into attacking the manufactured simulacrum while the cephalopod itself jets away to safety. If the predator gets too close, the viscous cloud coats its gills, smothering it; in the case of a moray eel, which sniffs out its prey, the ink inhibits its ability to smell.[26] Think of what happens to an inquisitive dog that startles a skunk.

Cephalopod ink is toxic, not just to prey and predators, but to the animals themselves. This presented a challenge to neurobiologist Bernd Ulrich Budelmann, former executive secretary of the Cephalopod International Advisory Council.[27] Conducting research for his doctorate, he used Volkswagens repeatedly to move scores of common Mediterranean octopuses (Octopus vulgaris) from the Max Planck Institute's satellite laboratory in Naples across the Alps to the main lab in Seewiesen, Germany, securing each animal in its own plastic bag to prevent it from inking its fellows to death. Sometimes Budelmann's octopuses became victims of their own defense mechanism. He told me that even confined to their separate quarters, about 10 percent of the animals inked themselves fatally when they became alarmed en route.[28]

Known to his friends and colleagues as "Ulli," Budelmann is a compact man with a gray beard and a wavy rumple of collar-length gray hair.

For 22 years he served as a research scientist with the Marine Biomedical Institute (MBI) at the University of Texas Medical Branch at Galveston. Although he retired in 2009, Budelmann remained captivated by cephalopods, especially octopuses.[29] "Most researchers who begin working with cephalopods never work with anything else," he told me. "The animals are so fascinating."

Using techniques ranging from neurochemistry to behavioral observation, Budelmann focused on the cephalopod central nervous system, especially equilibrium and vision. He operated on some of their brains and sensor organs to see what the effect would be on their ability to orient themselves. Most of his research involved the common octopus (*Octopus vulgaris*), a favorite among marine biologists, because the animals are active, inquisitive, and relatively easy to keep in captivity. However, making brain recordings in octopuses is difficult, so Budelmann implanted electrodes in live cuttlefishes and recorded their reactions.[30]

If this sounds cruel, consider that even humans have no pain receptors in their brains. The only thing that hurts when you have a headache is the membrane around your brain. Besides, Budelmann and his colleagues were studying how cephalopods behave normally, something animals in pain don't do, so he and his colleagues were careful to cause their subjects as little discomfort and alarm as possible.

The creature I saw flowing around the piling under the pier in Cozumel was a Caribbean reef octopus (*Octopus briareus*), which average less than half the size of Budelmann's research animals[31] but are otherwise similar. Full-grown, an average Caribbean reef octopus weighs about 2.2 pounds; its mantle measures about two feet across.[32] The one I spotted was probably a youngster, and its color shifts—from turquoise to lavender and back to turquoise—were probably due to sensing bigger animals (in this case two scuba divers) approaching it. An octopus registers every creature larger than itself as a potential threat. Changing colors and skin patterns (especially sporting huge mock eyes) may confuse a predator into perceiving the octopus as something too big to tangle with. Sometimes, octopuses flash bright colors in order to startle prey hiding motionless along the reef into moving.

Color can also indicate an octopus's emotional state. "By looking at their color and body texture, you can tell what kind of mood the animal is in," Budelmann said. "There are no other animals that allow you to see so quickly what is going on in the animal's brain."[33]

In other words, octopuses are the mood rings of the underwater world. In 1975, the height of the fad-mad era that brought the world hot pants and the Pet Rock, New York inventors Josh Reynolds and Maris Ambats hit on the idea of a ring whose "stone," a slice of quartz bonded to a liquid crystal, would respond to changes in the wearer's body temperature. Black allegedly indicated fear or depression; teal, optimism.[34] (Mood rings are still available online.)

Maybe the octopus I saw had been relaxed, noticed me and my buddy and became agitated, then calmed down as we began to swim away. Octopuses have a remarkable ability to change color and texture, which explains why that one was able to transform itself from turquoise to lavender to turquoise in seconds.

Like almost all other octopuses, Caribbean reef octopuses are solitary, spending their days in snug holes in the coral. Because their only hard parts are their parrot-like beaks, which they use to tear apart food, and the flexible cartilage encasing their brains, octopuses can squeeze into amazingly tight spaces. In captivity, they are famous for escaping their enclosures via pipes a couple of inches in diameter, only to be discovered the next morning as slimy blobs on the floor.

Even after they have reached their adult size, Caribbean reef octopuses change their lairs frequently. They avoid others of their species except during mating season, which on the Mesoamerican Barrier Reef lasts one to two months around January. After a male Caribbean reef octopus mounts a female, she can store his sperm for as much as 100 days while she looks for a suitable cranny in the coral to harbor her eggs—as many as 500. Unlike female squids, the octopus mother doesn't die immediately. Instead, during the 50 to 80 days it takes the eggs to incubate, she guards her clutch ferociously, not pausing to eat, not even consuming the occasional marauding crab she may kill as it tries to invade her nest. Shortly after the eggs hatch, she finally expires, typically of starvation.[35]

To make it to adulthood, an octopus has to be accomplished at both hunting and hiding. Out of hundreds of octopus eggs that hatch, only a tiny percentage of individuals survive the year it takes for them to reach sexual maturity. The vast majority die as hatchlings, dinner for juvenile fishes. As they get larger, octopuses move up the food chain, so that only sharks, rays, and big adult fishes, such as full-grown groupers—and people—will take them on.[36]

Scientists love mysteries, and octopuses and their cephalopod cousins offer them plenty. Cephalopod nervous systems are both like and unlike those of humans and our fellow vertebrates. On the "like" side, their complex nervous systems have similarities with those of vertebrates, so studying cephalopods can shed light on the human nervous system and brain. Squids and octopuses even show what neurobiologists call "omitted stimulus potentials"—brain functions that in human beings are associated with cognition.[37] This, combined with the animals' inquisitive behavior, has attracted the attention of psychologists and has led some researchers to speculate that cephalopods think.

On the "unlike" side, cephalopods function in ways that raise dozens of unanswered questions. For example, cephalopods' bundles of optic nerves connect their complex eyes to their optic lobes,[38] so how do a squid's eyes communicate information to its brain? And why can cephalopods change hues so quickly and dramatically yet be colorblind?

In his 2016 book *Other Minds: The Octopus, the Sea, and the Deep Origins of Consciousness*, philosopher Peter Godfrey-Smith posited that octopuses and their cousins evolved along a completely different track from vertebrates, with our last common ancestor a wormlike creature that died out 600 million years ago.[39] To thrive in their environments, Godfrey-Smith wrote, cephalopods "evolved large nervous systems, and the ability to behave in ways very different from other invertebrates. They did this on an entirely separate evolutionary path from ours."[40] Compared to humans, Godfrey-Smith explained, "Cephalopods are an *independent experiment* in the evolution of large brains and complex behavior."[41]

Unlike us, octopuses, squids, and cuttlefishes are masters of camouflage, but is this ability active or passive? Camouflage has three components: color, contrast, texture of the skin. As any scuba diver can attest, colors first shift toward blue and then disappear the deeper you go in the water. To see colors, you have to have more than one color pigment in your eye. Human beings have three. Almost all cephalopods have only one. Whether an octopus or squid is hunting or hiding, for camouflage the primary issue is matching the contrast pattern of something nonthreatening to prey or threatening to a predator. One thing the animals have is little reflector cells below their color cells, acting like mirrors.[42] Secreted in a patch of seagrass, for example, a squid might look like a few blades of vegetation.

Rather than camouflaging themselves by mimicking a background, which on a Caribbean coral reef could involve 600 to 1,000 species, each animal has a repertoire of a few optical tricks to confuse predators and prey.[43] Scientists have identified three basic types—uniform, mottle, and disruptive. Uniform and mottle tactics enable cephalopods to blend in with the background. By using the disruptive approach, they can disguise themselves as things that while clearly visible are of no interest to a potential predator—for example, a coconut shell. The relative simplicity of this system explains why cephalopods can transform themselves so fast—in less than a second.[44]

Cephalopods' brains and nerves are large enough to make them relatively easy to study. Also, being marine invertebrates, they are easier to keep in laboratories than, say, dogs or monkeys, and their use in experiments doesn't raise the unsettling sense that, if not for an accident of birth, Experimental Animal #456 might be Fido curled comfortably at our feet or that sweet simian mother endearingly nursing her baby in the local zoo.

It doesn't take a doctorate in marine biology to be fascinated by You-Tube videos of octopuses unscrewing jar lids to reach a small crab, differentiating shapes (a cross versus a circle or a triangle) to get to a yummy prawn on the other side of a swinging door, even learning by watching another octopus. Octopuses can learn through touch, something that squids can't do, and are famous for their short- and long-term memories.[45] Surely, these animals must be highly intelligent.

They even *look* smart, at least to laypeople. An octopus's floppy mantle, which resembles an unstarched chef's toque, could easily be mistaken for a large head. Its big eyes make the animal appear downright brainy, as if a pair of horn-rimmed glasses would be the perfect accessory.

However, Budelmann cautioned against making too much of brain size, let alone brainy appearance. "It's dangerous to conclude that cephalopods are superintelligent because they have large brains," he said. "Brain size alone doesn't tell you that much about what the animal can do. Neither does brain-to-body weight ratio. The number of nerve cells is more indicative. The more units you have, the more complex interactions you have. Some invertebrates have large brains but not many components."[46]

Like all animals, including humans, octopuses possess sufficient intelligence to secure food and to avoid *becoming* food long enough to reproduce. Without question, octopuses are smart, but smart compared to *what?* More cautious marine biologists put them at the top of the invertebrate heap, comparing them to fishes and even amphibians. On the other end of the scale, naturalist Sy Montgomery, in *The Soul of an Octopus,* her charming book describing her personal experiences with octopuses in the New England Aquarium in Boston and in the wild off Cozumel and the French Polynesian island of Mooréa, concluded that they have consciousness. She described sitting in a church in Mooréa, listening to a service in Tahitian, a language she didn't understand, and thinking, "I am sure of one thing as I sit in my pew: If I have a soul—and I think I do—an octopus has a soul, too."[47]

But does such cross-species empathy reflect objective reality, or does it reflect something in our own emotional makeup? Is it about the animal, or is it about us?

Both individually and as a species, human beings crave company. That's why we send out probes searching for intelligent life on other planets. Partly as a result of this yearning, we anthropomorphize other animals. On the level of basic emotions, such as fear or maternal attachment, this may not be off-base. When a rabbit startles at the sight of a coyote or a guinea pig snuggles down with her pups, their endocrine systems exudes the same chemicals that surge in a man surprised by a mugger or a mom cuddling her baby.

Between my first and second years of graduate school, my then-husband Phil and I spent the summer on Oahu. At the time, which was the height of the Vietnam War, the federal government had a program to provide temporary replacements so that permanent civilian staff, GS-4 and below, could take vacations. Applicants could choose among a variety of locations, ranging from San Antonio, Texas, to Frankfurt, Germany. We picked Hawaii, where I snagged an assignment as a supply clerk at Hickam Air Force Base and Phil participated in a Department of Commerce project testing the water around schools of tuna.

In the huge converted hangar that served as Hickam's center for supplying the troops in Southeast Asia, I befriended a young lieutenant, who invited Phil and me to dinner. We began spending most Friday evenings at the cottage the lieutenant shared with his wife, playing canasta and drinking San Miguel beer. In those days, San Miguel lacked quality control, so each bottle was an adventure, sometimes wildly different from its predecessor in alcohol content, taste, and amount and texture of sediment.

The most curious aspect of these pleasant visits was the opportunity to observe our hosts' pets, a medium-haired gray cat and a young shepherd-mix dog, both females. Although she was probably nearing the end of her reproductive life, the cat had recently given birth to a litter of kittens. The dog had yet to come into her first heat, yet the dog provided most of the mothering. Right after nursing, the cat would spring from the cardboard box containing her litter of kittens and meow to be let out. Meanwhile, the dog would lie down near the newborns, her ears peaked at half-alert. Whenever the kittens mewled, the dog would spring to their box, woof at them softly, and lick them gently. The cat returned only when her teats were again full.

This maternal display provided for great amusement and speculation on the part of us humans. It seemed obvious that the mother cat had conned the young bitch into taking responsibility for her offspring, so that she could be free to come and go as she pleased, like a human mother hiring a teenage babysitter. Now, as I think back on the scenario, I recognize that the four of us were anthropomorphizing like mad, attributing human motivation and conscious manipulation to animals based on casual observation. Given that cats and dogs are much closer to us on the evolutionary ladder, we make a far greater leap when we speculate that an octopus must be lonely or bored or even feel affection for us and dislike for another person.

That doesn't mean that we're wrong to do so. It just means that scientifically, we're over very deep water. We will have a lot of meticulous work to do if we are to explain how, for example, hormones that we may share with a mollusc can be processed in its brain—or in its thousands of suckers—into anything like our own experience. Granted, human emotion involves chemically triggered sensations that we may well have in common with all other sentient animals, but it also involves cognitive interpretation.

"As a serious scientist, you should describe a phenomenon but avoid putting too much time into interpretation," cautioned Budelmann. "Octopuses are fascinating animals, and we love them. We want them to do all kinds of things. That's why we need to beware of anthropomorphication."[48]

Take boredom. Montgomery mentions that several of the octopuses she observed in the New England Aquarium appeared to be bored and speculated that boredom may have prompted one to escape overnight from her tank (with fatal results, since octopuses can't breathe air). In the wild, octopuses live in environments rich in sensory stimuli. Tropical coral reefs are very complex, and even the rocky coastal waters off Alaska have a lot going on. To thrive, or even survive, an octopus needs to explore its surroundings. Then humans catch it and put it in a relatively simple aquarium, and what is the animal supposed to do? It doesn't interact with its own species, except once in its life, when mating. But out on the Mesoamerican Barrier Reef, for example, it does feel and taste the hard corals, the gorgonians, and the crabs and prawns it favors as foods. Deprived of all this accustomed stimulation, any octopus worth its saltwater will explore its enclosure and the hands and arms of the attendants who feed it. It will even probe and bounce pieces of PVS tubing and acrylic cubes and stack Legos placed in its tank.

"The animals are very inquisitive," Budelmann explained. "They manipulate things, but the results (like opening a jar) may be accidental."[49]

No matter how much we may want to, we can never know what, and how, an octopus experiences the world around it, how it feels or what it thinks, assuming that it *does* think. Consider that we often misinterpret observable emotional cues from members of our own species, even the people closest to us, and we can never feel someone else's pain or think another person's thoughts. Empathy has its limits.

Octopuses needn't be like us to fascinate us and to be worthy of our protection. By observing them—in laboratories, in aquariums, and in the wild—and refraining from drawing anthropomorphic conclusions, we can gain insight into the planet we share and learn to value its diversity.

Like all the creatures that live along the Mesoamerican Barrier Reef, octopuses and their cousins face threats from pollution, from unsustainable fishing (not just of them, but also of the species they hunt and eat), and from rising temperatures and increasing acidity of the water that bathes the coral that forms the foundation of their environment. Being high up the food chain has its advantages, but it also renders these cunning creatures dependent for their survival on the well-being of all the species below them.

The thrill of encountering an octopus as it flows along the reef at night, like the thrill we feel watching a shoal of squids swimming in exquisite precision, comes because these animals are so different from us, not because they are similar. If they were to disappear, a measure of the world's wonder would vanish with them.

13

AQUATIC INTELLECTUALS

Dolphins

As fascinating as I find the hard and soft corals, the sponges, and the critters on the reef itself, when I dive along the wall of the Mesoamerican Barrier Reef, I try to turn to the other side every so often and look out at the blue expanse extending seaward. The drop-off might be a hundred feet, or it might be a thousand; but this is where the big creatures are—manta rays with 23-foot wing spans, whale sharks longer than my living room, schools of tiger sharks, pods of dolphins. I won't see marlins, swordfishes, tunas or mahi-mahis. These hang out far from shore, over the depths, which is why trophy anglers call their sport deep-sea fishing.[1] Nor will I see true whales, which avoid this quadrant of the Caribbean altogether, instead migrating along the Bahamas and the Antilles, 800 miles to the east.

What I do see, when I'm lucky, are some of the smartest animals on the planet, animals that also happen to be some of the most lethally efficient hunters. They don't live *on* the Mesoamerican Barrier Reef, but they live *with* it. They perform a vital role in the ecosystem, collectively controlling populations ranging from squids to full-sized groupers. Like porpoises and whales, dolphins belong to a group of aquatic animals called cetaceans (order: Cetacea). All cetaceans share a common ancestor with hippopotamuses. After evolving on land, some of the descendants of this creature returned to the sea about 55 million years ago but continued to breathe air.[2]

Comedian Will Rogers quipped famously: "If dogs don't go to heaven, then I want to go where they go." I feel similarly about dolphins. These remarkably intelligent, social, and even playful marine mammals make me happy that we inhabit the same planet. However smart octopuses may be—and whatever way they may be smart—dolphin intelligence is unquestioned. Just for starters, they have a centralized brain, a single organ, not like octopuses that have a far-flung system of sensors, whole sections of which operate independently. And pound for pound, the ratio of a dolphin's brain to its body mass is larger than that of a chimpanzee's, our closest relative.[3]

Worldwide, the common bottlenose (*Tursiops truncatus*) is the most prevalent of the 32 species of dolphins and also the species most frequently sighted along the Mesoamerican Barrier Reef. Atlantic spotted dolphins (*Stenella frontalis*), spinner dolphins (*Stenella longirostris*), and orcas (*Orcinus orca*), which despite their popular name "killer whales" are not whales, but the largest dolphins, make occasional appearances, but bottlenose dolphins are regulars. Seven to 13 feet long and ranging in color from medium gray to near black, they sport a distinctive long beak, or rostrum, and a scimitar-shaped dorsal fin. Casual observers sometimes confuse dolphins with porpoises; but porpoises have shorter, stockier bodies, blunter heads, and triangular dorsal fins. Their teeth, as well, have different shapes—dolphins' choppers are conical; porpoises,' spade-shaped. (Both use their teeth to catch and kill, but not to chew, their food, instead swallowing a fish whole headfirst—to avoid being stuck by its spines.) Also, while bottlenose dolphins favor temperate and tropical waters around the world, ranging in the Western Atlantic from New Jersey to southern Argentina, porpoises prefer cooler, even cold, water and eschew the Caribbean.[4]

And unlike dolphins, porpoises don't whistle. One of the ways dolphins communicate with each other is by whistling through their blowholes. They also emit clicks and occasional loud bursts of sound, which they use to ward off sharks and to discipline young. Dolphins communicate through body language, as well.[5]

As a juvenile, each dolphin adopts a series of whistles that researchers believe functions as its name—like answering the phone "Joe here" when the dolphin uses its own whistle pattern or saying "Hi, Karen" when one dolphin recognizes another.[6] And they do seem to recognize acquaintances, even when they haven't seen each other for 20 years.[7]

In another distinctive feat of recognition, dolphins, like humans, can recognize themselves in a mirror. Most other animals act as if they are encountering another individual when they see their own reflection.[8]

During the years I lived on Galveston Island, bottlenose dolphins were such a common sight that locals pretty much ignored them. When friends came down from Houston, I would take them for lunch at a restaurant on the harbor and urge them to sit facing the water.

Pretty soon they'd exclaim, "Look! A dolphin! No, two! No, five!"

"It's a pod," I'd explain calmly. "That's what you call a group of dolphins.

They can have up to 30 members, but around here it's more like 10 or 12. One pod lives in the harbor. A few others live in the bay or just offshore."

Although I wouldn't allow myself to express as much excitement as my guests, I would still turn around to watch as the dolphins arced gracefully over the surface, the sun hitting their backswept dorsal fins. This spectacle, not the $13 fish tacos, was the reason we had chosen this somewhat pricey venue, after all.

Like all cetaceans, dolphins are mammals. They bear their young live, almost always singly, very occasionally as twins. A bottlenose dolphin pregnancy lasts 12 months. The newborn calf emerges tail-first, preventing it from drowning during the hours-long birthing process. In addition to producing high-fat milk to nurse their calves, dolphin mothers safeguard them and teach them how to find their own food and interact with other dolphins. Mother dolphins even show their daughters (but not their sons) how to use sponges to protect their beaks when they burrow for food in the sand—an example of tool use. Biologists call passing along knowledge to other individuals "culture."[9]

Dolphins have culture.

Bottlenose dolphins mature sexually at about nine and have a natural life span around 50 years.[10] Because calves usually nurse for two or three years and moms always wean one calf before starting another, a female dolphin might reproduce only every four, five or even six years, but the care she invests in her young results in a high survival rate.[11] About 97 percent of bottlenose dolphin calves survive their first year,[12] compared to a bit less than 96 percent of human babies.[13]

Once weaned, juvenile dolphins stay close to their mothers for a few years, learning the skills of hunting and social life.[14] Eventually they leave her side to hang out with other adolescent dolphins of the same sex. Males often form tight, even lifelong, bonds with their peers. Sometimes a male pod, acting together, will separate females from the main pod and treat them like a harem, giving the dominant male in the posse breeding privileges.[15]

Even more than people, dolphins are social creatures. They live in groups. They cooperate to chase off their sole nonhuman predators—large sharks. When a dolphin becomes ill, the fellow members of its pod will support it, so that it can stay near the surface and breathe.[16] If one dolphin heads for shallow water, the other members of the pod may follow, resulting in a tragic mass stranding.[17] Generally, however, the ability to cooperate efficiently on complex tasks benefits the survival of the group and its individual members.

Dolphins often hunt collectively, forming a circle to herd prey fishes into a bait ball or a line to drive them toward the shore.[18] Scientists observing dolphins in Florida Bay have watched as a pod stirred up the muddy bottom to form an aquatic corral, then moved back a few yards to catch the panicked fishes as they jumped over the barrier.[19]

Using a specialized natural sonar called echolocation, dolphins can also hunt effectively alone. Beaming sound waves through a fatty organ in their foreheads called a melon, then registering, via sensors along their heads, the waves that bounce back, a dolphin can discern the size and shape of a fish a mile away,[20] then dash as fast as 20 miles an hour to snatch it.[21] Sound travels four times faster underwater than it does in air.[22]

Dolphins even collaborate to help humans. In addition to pods protecting swimmers from sharks or rescuing distressed divers, in one coastal community in Brazil, dolphins regularly cooperate with local fishermen to their mutual advantage. Mullet form a staple in the diets of both, but the water is so murky that the humans need help to see where the fishes are running. Using their biosonar, the dolphins, however, can tell. When they spot a school of mullets, they herd it toward shore, slapping their tails on the water, showing the fishermen where to cast their nets. The fishermen enjoy abundant catches, and the dolphins feast freely on the confused mullets that jump around trying to escape. Locals have told anthropologists that this mutually advantageous tradition dates back centuries.[23]

More controversial instances of dolphin-human cooperation involve military purposes. During the 1970s, the U.S. Navy (and reportedly the Soviet navy as well) began training and using dolphins to locate and tag underwater mines and to identify enemy divers and unusual objects, such as spy submarines.[24] Even when visibility approaches zero, dolphins can use their biosonar to tell whether an object is wood, metal or plastic from a hundred feet away.[25] Despite the objections of animal rights activists and the availability since 2014 of underwater robots to serve similar functions, the military still uses dolphins, apparently because they perform better.[26]

Although the beak that inspired the name "bottlenose" may look like a nose, it has no respiratory or olfactory function. The limited sense of smell a dolphin possesses comes via the blowhole (its real nose) in the top of its head, and that's closed when the animal is underwater. Bottlenose dolphins breathe three or four times a minute, even when they're asleep. Unlike us, they don't breathe automatically, so they turn off one hemisphere of their brains when they're sleeping, leaving the other hemisphere to manage surfacing and breathing.[27] Like us, they sleep about 8 hours out of every 24, a pattern that for them gives each side of the brain 4 hours of rest; but unlike us, they snooze for no more than a few hours at a time.[28]

When dolphins sleep, they close one eye, the one on the opposite side of the brain from the hemisphere that's resting.[29] Set on the sides of their heads, their eyes operate independently. With their visual cortex located next to their auditory cortex, dolphins can interpret images as sounds and sounds as images.[30] Dolphins have good eyesight, even in murky water, but they lack binocular vision, the ability that we, and many other creatures of the land and air, share, allowing us to judge an object's distance by focusing on it using two eyes set a few inches apart.

During four decades of living in and visiting Galveston, I never tired of watching dolphins. When I ran along the seawall, I sometimes caught sight of a pod 100 yards offshore. When I took the car ferry across to the Bolivar Peninsula, bottlenose dolphins often followed the boat. Maybe they were feasting on fish disoriented by the churning motors. Or maybe they were just playing. Decades earlier at the beach in Ocean City, Maryland, I saw the distinctive dorsal fins of a pod of dolphins as they approached a group of surfers sitting on their boards awaiting the next wave. One of the dolphins swam under a surfer and rose abruptly, knocking him over. The culprit surfaced, exhaling forcefully and opening its mouth. I could swear that that dolphin was laughing.

Episodes one and three of the 2017 BBC documentary series *Blue Planet II* showed bottlenose dolphins riding the surf off the coast of South Africa. "Could they be playing?" narrator David Attenborough speculated, a touch of awe flavoring his voice.[31]

Decades earlier, researchers had settled the question that dolphins play. They do, as do other social marine mammals (and a number of terrestrial mammals, as well). At the Kaikoura seal sanctuary along the east coast of New Zealand's South Island, I watched as, with several adult females looking on, eight juvenile seals took turns on a natural rock slide formed where the waves broke. After sliding down barking animatedly, each young seal would climb back up the wet boulder and take its place in line. These were not trained seals. Their play seemed completely natural.

Play helps dolphin calves learn how to behave socially, as well as enabling them to develop skills useful in hunting and mating.[32] Adult dolphins play, too, and even imitate other species, including humans; researchers have observed them tossing around clumps of sargassum seaweed like toys[33] and adorning themselves with leaves.[34] But juveniles are more innovative and creative.[35] Play also helps young dolphins develop spontaneous problem-solving skills, something at which they excel.[36]

Given the clear willingness of dolphins, even in the wild, to interact with humans and the creatures' use of sound and gesture to coordinate pod activities and protect and instruct their young, it was logical, maybe even inevitable, that people would try to take communication with dolphins beyond the level of tail-slaps signaling Brazilian fishermen to cast their nets. What if those patterns of clicks were a language, or at least a vocabulary of nouns? If one set of clicks was analogous to a dolphin's name, couldn't another denote "school of mullets" and another "sargassum"?

Among the first to pursue this intriguing question with scientific rigor was U.S. National Institute of Mental Health neurophysiologist John C. Lilly. Lilly began studying dolphins in the 1950s, then rode the wave of the counterculture through the 1960s and beyond, until his death in 2001. Written for lay audiences, his books on dolphin intelligence became best sellers, especially appealing to people replacing conventional religious beliefs with New Age mysticism. Lilly even injected some of his dolphin subjects with LSD at the research facility he established in the U.S. Virgin Islands,[37] a partially flooded house designed for human-dolphin cohabitation.[38] Convinced that dolphins possessed their own language, he tried unsuccessfully to teach one of them English.[39]

Despite his name and his interest in drugs, John Lilly was not a member of the Eli Lilly pharmaceutical dynasty. However, he did come from money. His father was a bank president in Saint Paul, and his mother's family owned stockyards there.[40] Lilly's father provided part of the funding for his research. In addition to being an MD and a psychoanalyst, Lilly was the inventor of the isolation tank, which employed sensory deprivation as a means to explore human consciousness. With a history of experimenting on himself that dated back to his days in medical school, Lilly would drop acid while floating in an isolation tank suspended above a pool of dolphins.[41] His friends included countercultural icons Timothy Leary, Ram Dass (born Richard Alpert), and Allen Ginsberg.[42] Lilly went on to pioneer attempts to make contact with extraterrestrials, but he never succeeded in bridging the language barrier between dolphins and humans. His unconventional research methods, combined with his wild-eyed zeal, may have set back serious investigation of the topic more than advancing it.

It took another generation of lower-key researchers to return scientific respectability, and major funding, to the exploration of human-dolphin communication. Comparative psychologist Stan A. Kuczaj succeeded in demonstrating that dolphins at least communicate with each other on a sophisticated level. Working at the Roatán Institute for Marine Sciences, a combination research center and resort at the southern end of the Meso-american Barrier Reef, Kuczaj used a video camera equipped with an underwater microphone to document a pair of dolphins as they received the hand signal for "innovate" followed by one for "tandem." Sinking beneath the surface, the two 400-pound marine mammals exchanged a series of chirps, then rolled over in unison and delivered three tail flips, like a pair of giant synchronized swimmers. Rising seven more times to receive the same signals, the dolphins proceeded to exhibit a different precisely coordinated set of behaviors each time, including strutting on their tails, blowing bubbles and executing a pirouette.[43] This was not a routine humans had taught them. Either the pair were utilizing those chirps to decide what trick to perform next or something even farther out—say, dolphin mental telepathy—was going on.

Kuczaj wrote his doctoral dissertation on the development of language in young children, particularly the understanding of such verb forms as gerunds and participles. He went on to blend comparative psychology, behavioral science, and marine mammalogy. Although he described himself as an ex-hippie and once owned two Volkswagen vans and traveled in Nepal, his research, which resulted in some 150 publications, stayed within the bounds of accepted scientific method. When Kuczaj died unexpectedly in 2016 at the age of 65, he was a full professor at the University of Southern Mississippi and director of the Marine Mammal Behavior and Cognition Laboratory.[44] He had earned, and maintained, his academic chops.

Unlike Kuczaj, Lilly, and most other dolphin researchers, behavioral biologist and founder of the Wild Dolphin Project Denise Herzing was studying dolphins in the wild. Focusing on Atlantic spotted dolphins, smaller than bottlenose dolphins but so closely related that they sometimes mate, Herzing was endeavoring to learn the animal's language,

rather than trying to teach them ours. Reasoning that if dolphins identify themselves, and each other, through distinctive patterns of clicks or
whistles, they might denote objects in their environment in a similar way,
Herzing partnered with Georgia Tech computer scientist Thad Starner to
develop a device to translate human to dolphin language and vice versa.
The Cetacean Hearing and Telemetry (CHAT) water-tight computer
transmits prerecorded dolphin whistles and records the animals' whistled
replies.[45]

At the time this book went to press, Herzing was attempting to get a
group of three young female dolphins to associate each of three prerecorded whistle patterns with three objects—a rope, a piece of sargassum,
and a red scarf. She carried the scarf in her bathing suit, then pulled it out,
simultaneously playing the designated series of clicks from the 20-pound
CHAT strapped to her chest. Those three words could form the basis of a
shared artificial language.[46]

Although that language would be artificial, it would still constitute an
astounding breakthrough in interspecies communication. American Sign
Language is artificial, and look what worlds it has opened up. Perhaps
more intriguing is the algorithm Starner developed that sorts a jumble of
images or sounds into meaningful patterns. As Starner told Joshua Foer of
National Geographic, the algorithm has already identified consistent signature whistles that mother dolphins and their calves use with each other
and suggests that these might be combined in some significant way.[47]

Using their own signature whistles, Herzing taught two young wild female dolphins to appear beside her boat; but whether their interaction
had a linguistic component is questionable. When Herzing presented the
red scarf and played the "red scarf" series of whistles on her CHAT, the
dolphins played with the object, passing it back and forth and waving it
around. Eventually they returned it to Herzing, but they never used the
human-devised "red scarf" whistle series.[48]

Human speech is a remarkably complex endeavor. It requires vocal
chords; a particular shape of the jaw, tongue, and palate; the ability to
control breath; and a brain capable of interpreting information beyond
denoting—or pointing. (Although an exception to some of these require-
ments, sign language is based on spoken language.) Despite our huge and
flexible vocabularies, we humans often fall back on gestures to communicate. We shake an index finger when we disapprove, shrug our shoulders
when we're baffled, flip the bird to a fellow driver who cuts us off.

The scientists who raised a female mountain gorilla from birth succeeded in teaching her to "speak" a version of American Sign Language at the level of a young child's mastery of spoken language. Koko was even able to craft insults, once calling a person who irritated her "a toilet."[49]

The prospect of humans "speaking" with dolphins faces significant physiological hurdles. For one thing, dolphins lack vocal chords. Granted, some human languages rely heavily on clicks, but they also involve sounds produced in the voice box. For another, unlike Koko, the gorilla that learned a modified version of American Sign Language, dolphins lack hands and fingers.[50]

Modern humans (*Homo sapiens*) and chimpanzees (*Pan troglodytes*), our closest evolutionary kin, last shared a common ancestor sometime between 13 and 10 million years ago; with great apes like Koko, 17 million years.[51] We parted evolutionary ways with the ancestors of dolphins more than 50 million years ago.[52]

In proportion to our body weight, our brains are three times larger than those of other primates. Much of that difference involves our big neocortex, which in humans is responsible for language, self-awareness, problem-solving, and abstract thinking,[53] including our ability to imagine things like money and institutions into existence.[54] Yet while dolphins have a smaller cortex in relation to their overall brain size, they are quick problem solvers—and given their signature whistles and complex social lives, they are clearly self-aware. Maybe they use a different part of their brains to solve problems, or maybe their cortexes are more efficient than ours. Their cortical ridges and grooves are deeper, allowing greater blood flow.[55]

But dolphins clearly experience the world in ways very different from the ways we do. Take living underwater but breathing air and having to surface to take conscious breaths—or having an auditory nerve twice the size of ours.[56] Although dolphins can see, they must relate to their environment primarily through sound. We relate to ours primarily through light.

Both humans and dolphins have invested a huge amount of evolutionary capital in developing big brains, but our brains are so different in structure, so well-adapted to our survival in our own environments, that our attempts to either translate "dolphin" into "human" or create a common artificial language may be, literally, whistling in the dark.

14

MISUNDERSTOOD
"BAD BOYS OF THE REEF"

Sharks

As much as we may feel a warm attachment to dolphins, and may even identify with them, many people feel repelled by sharks. Simply catching a glimpse of a Caribbean reef shark (*Carcharhinus perezii*) minding its own business, focused on finding some toothsome fish, is enough to cause a diver to "suck air," inhaling and exhaling so rapidly that his or her tank becomes depleted and a potential 45-minute dive gets cut short at 30.

Why do we find sharks frightening? Is it just the Hollywood hype, or is there something atavistic behind our anxiety? Probably a bit of both.

There are the movies—not just the *Jaws* series but others that depict, for example, sailors knocked overboard being quickly surrounded by a school of ravenous sharks. But these thrillers exploit far more ancient anxieties. Our early ancestors learned to fear animals that could eat them, and we still recognize that being eaten would be an exceptionally terrifying and painful way to die. Also, we fear losing control, some of which we relinquish simply by getting into the water, which isn't our natural element. Having an animal attack or even threaten us delivers another blow to our sense of command.[1]

Sharks also scare us because they look like relics from a much earlier era. That's because they are. The first sharklike creature appeared in the world's oceans 450 million years ago—before there were trees. Although they evolved significantly, the basic design—a fish with a cartilage skeleton that could move quickly and flexibly through the water and, eventually, use its sharp, bony teeth to catch and tear into prey—was so successful that sharks survived five mass extinctions, one of which killed off 96 percent of all marine life. Because they survived, sharks dominated the ancient seas, and they maintained that domination over the eons.[2]

Another reason sharks are frightening is all those teeth. Great white sharks (*Carcharodon carcharias*) have 300 sharp teeth arrayed in several rows, but the arrangement isn't just for predatory purposes. Shark teeth are made of dentin, which is harder than bone, but because they are anchored in comparatively soft cartilage, they can get knocked out easily, even in the process of catching prey. Sharks can regrow their teeth, but that takes time, so having spares on hand is useful.[3]

Despite our fears, humans are not on sharks' diet. We are creatures of the land; they are creatures of the sea. Although some terrestrial animals, such as bears, may deliver fatal attacks, the list of predators that actually like to eat us is relatively short: leopards, lions, tigers, Komodo dragons, crocodiles, and alligators.[4]

The world's 465 species of sharks range in size from the seven-inch pygmy (*Euprotomicrus bispinatus*) to the 40-foot-plus whale shark (*Rhincodon typus*).[5] Unlike humans, sharks continue to grow as adults.[6]

Observers have identified 30 shark species on the Mesoamerican Barrier Reef.[7] Some sharks feed on plankton and krill; most, depending on where they live, like fishes, dolphins, seals, and other sharks. When they attack humans, sharks are either confused—for example, mistaking a swimmer for a floundering snapper or a glittering silver ankle bracelet for the scales of a mullet—or involved in a feeding frenzy, hitting a ball of baitfish and biting a nearby swimmer in the process.[8]

Thirty-two percent of shark attacks are on swimmers; 65 percent are on surfers. Scuba divers and snorkelers face little risk.[9] But even relatively placid sharks can bite if they feel threatened. That's why when entering a swim-through tunnel in the reef, I always take a peek inside. Once, I encountered an eight-foot-long nurse shark (*Ginglymostoma cirratum*) and had to back paddle. Clearly awoken from a nap, it eyed me with obvious annoyance.

Humans are far more dangerous to sharks than the other way around. We kill about 100 million sharks a year, some inadvertently as bycatch, others intentionally, sometimes through the cruel practice of removing their fins for soup and throwing the now-helpless and bleeding shark back into the water to die.[10] By comparison, sharks attack around 80 to 130 people a year worldwide. Most of the resulting injuries are not fatal.[11]

As the human population has swelled and beaches have risen in popularity and accessibility, the number of shark attacks has increased. But an individual's chances of being bitten by a shark have actually declined by 91 percent since the mid-1950s. Established in 1958 and housed in the Florida

Museum of Natural History at the University of Florida in Gainesville, the International Shark Attack File has logged 6,500 shark attacks on humans, but this number includes historical reports dating back to the sixteenth century.[12] Today, a person's chances of being killed by a shark are one in 3.7 million.[13] An individual is far more likely to be done in by a falling icicle than by a shark.[14] In 2015 and 2016 more people died while trying to take selfies than died from shark attacks.[15]

By comparison, a person's chances of dying from drowning are one in 1,113. That's the real danger in the water.[16]

In 2018, 130 people were involved in shark attacks worldwide. Thirty-four of those attacks were provoked. One was in a public aquarium. Nine attacks were directed at boat hulls or motors.[17] Worldwide, an average of 10 people are killed by sharks annually, and only 30 species of sharks have been involved in reported attacks on people.[18] White (*Carcharodon carcharias*), tiger (*Galeocerdo cuvier*), and bull (*Carcharhinus leucas*) sharks are responsible for most of the attacks on humans.[19]

Because a surfboard is roughly the size and shape of a seal floating on the surface, surfers suffer almost two-thirds of shark attacks on humans, and white sharks are responsible for a disproportionate number of those. And because this particular shark tries to make its first strike a mortal blow, even nonfatal attacks often result in serious injuries.[20]

Visitors to the Mesoamerican Barrier Reef have nothing to fear from white sharks. Although they are sighted with some regularity off Florida, white sharks prefer temperate, rather than tropical, waters and prey on seals, which like things cool. Like all other fishes, white sharks are cold-blooded. This means that the animal's body temperature rises and falls with the temperature of its surroundings—in the case of fishes, water. But the white shark can raise the temperature in its stomach as much as 57°F above the water temperature.[21]

The most dangerous shark species on the Mesoamerican Barrier Reef are tiger and bull sharks, but shark attacks are rare in the Caribbean. Tiger sharks can grow to over 20 feet long, so even the casual flip of one's tail can cause injury.[22] They are fiercely territorial and will chase both other sharks and humans off their stretch of reef.[23] Voracious eaters, tiger sharks have been known to consume license plates and old tires. Unlike great whites, which have discerning palates and will often spit out a surfer after one taste of a neoprene wetsuit, tiger sharks are inclined to stick around and finish the meal.[24]

To locate prey, tiger sharks rely mainly on their sense of smell and their hearing, because they hunt long-range. Once they get close to their victim, vision takes over.[25] When attacking, they can generate speeds up to 19 miles an hour—heart-stopping to watch, though sluggish compared to velocities attained by shortfin mako sharks (*Isurus oxyrinchus*), which have been clocked at 41. (Speedo studied the mako's skin in developing fabric for a new suit targeted at competitive swimmers.) Although they inhabit warm waters around the world, makos don't frequent the Mesoamerican Barrier Reef, seeming to prefer the North Atlantic's Gulf Stream.[26]

Bull sharks present a hazard to humans because they hang out in shallow coastal waters, which are often murky. Almost uniquely among sharks, they can tolerate brackish and even freshwater. They rely on hearing, smell, and their lateral lines to locate prey.[27] Bull sharks can easily mistake swimmers for a school of fishes and can be drawn to the scent of the catch when fishermen cast their nets in estuaries. And bull sharks breed along the Mesoamerican Barrier Reef, in the lagoons off Playa del Carmen, Quintana Roo, Mexico.[28]

The most common sharks along the Mesoamerican Barrier Reef are Caribbean reef sharks (*Carcharhinus perezii*),[29] which tend to ignore humans unless they're spear fishers trailing their bloody catches on stringers. One of the handsomest residents of the reef is the lemon shark (*Negaprion brevirostris*), distinctive because of its yellowish back and sides.[30] Another regular denizen, the black-tip reef shark (*Carcharhinus melanopterus*),[31] is exceptionally photogenic because its fins show up so well.

Sharks have five different kinds of fins: pectoral, dorsal (that distinctive triangular fin on their backs), pelvic, anal, and caudal (the large, upswept fin at the end of their tails).[32] Unlike those of bony fishes, which flex, sharks' fins are rigid.[33] Located on each side of the shark just below and slightly behind its gills, the pectoral fins provide lift, like wings on a plane. The pelvic fins lend stability as the shark swims. The dorsal fin (either a single one or a big one with a smaller one farther back) also helps the animal stay stable. Located on the lower part of the body, the anal fins aid stability as well. The caudal fin propels the shark through the water.[34]

Although their fins are standard-issue, the perpendicular cranial design that inspired their name makes hammerhead sharks (family: Sphyrnidae) among the most bizarre creatures on the reef. They look like a child has cobbled them together from several animal action-figure sets, attaching the T-bar of the head onto a shark's body, then sticking a couple of bulging eyes on the ends. The most memorable time I observed hammerheads in the wild was as I began my ascent from my first dive into Belize's Blue Hole, the iconic sinkhole in Lighthouse Reef. Looking up from 125 feet down, I saw six hammerhead sharks circling a few yards beneath the surface, outlandish head to familiar shark tail. That sight sometimes appears in my dreams, but never in my nightmares. Even my subconscious seems to know that hammerheads rarely attack humans.[35]

Hammerheads are the most recent evolutionary iteration of the shark design. That mallet-shaped head gives them four important advantages as hunters. First having their eyes set far apart and facing forward provides depth perception. Not only can hammerheads spot prey, they can tell how far to lunge.[36] By swinging their heads, these highly advanced predators can even achieve panoramic vision.[37] Second, that perpendicular bar sports an array of nostrils, which, although they have no breathing function, endow the hammerhead with a keen sense of smell. Third, that broad head supports a line of electroreceptors, called the ampulla of Lorenzini, which can detect the weak electrical field that all animals emit when they contract their muscles.[38] Thanks to these electroreceptors arrayed on the fronts of their heads, all sharks can "see" in the dark[39] and use electroreception to locate prey they couldn't find with their other senses. But a hammerhead has these sensory organs arrayed along a proportionally wider area, making them especially effective, for example, at locating a stingray buried in the sand.[40] Furthermore, the hammerhead shark can use its broad head to hold that ray down while consuming it. The hammerhead's mouth is located conveniently on its underside, a couple of feet behind that distinctive bar bristling with sensory organs.[41] That idiosyncratic head shape may also account for the hammerhead's remarkable ability to navigate using the magnetic fields produced by seamounts.[42]

Like many other varieties of carnivorous sharks, hammerheads have a natural set of goggles to protect their eyes, especially when the shark is attacking an animal that fights back by thrashing around with its fins. Called a nictitating membrane, this milky-hued, almost opaque inner lid makes the shark look even more terrifying—otherworldly, like a zombie.[43]

Except for rays, skates, and chimaeras (not to be confused with "chimera," the mythical Greek monster), sharks are the only fishes with skeletons made of cartilage, rather than bone. Covering the cartilage is a thin calcified layer,[44] similar to what humans have separating cartilage from bone in our joints. Because cartilage is much lighter than bone, sharks don't require the gas-filled swim bladders that help bony fishes control their depth. Also, sharks carry a lot of low-density oil in their livers, and this provides some buoyancy—although not enough; so to prevent sinking, sharks need to move continuously. Some species, like the nurse shark, solve the problem by finding caves or tunnels on the reef, where they can rest safely on the bottom.[45]

All fishes breathe through their gills, most with five or fewer pairs, which are protected by a covering. Sharks have between five and seven pairs, set unprotected along their sides. Even sharks' skin differs from that of other fish: Their scales are formed of the same material as their teeth, and they point upward, imparting a texture so rough that sharkskin has been used as sandpaper.[46]

Sharks are "apex predators"—at the top of the food chain, with no natural enemies except other sharks and us.[47] Scientists used to think that sharks could live 20 to 30 years in the wild,[48] but more recent research dating carbon from radioisotopes absorbed by various species of sharks, including tiger sharks, in the form of fallout from nuclear bomb tests of the 1950s and 1960s suggests that some sharks can live for a century or more,[49] performing the essential ecological function of keeping other species in check.[50]

Yet, in parts of the world where shark meat constitutes a dietary staple, the populations of some species of sharks dropped 75 percent between 1990 and 2010. The most obvious reason is the toll taken by a growing human population, not only for sharks specifically but for all kinds of fishes. Sharks and fishermen pursue the same prey, and a shark bent on hunting can easily find itself snared in a net filled with anchovies.[51]

But another reason is that sharks have very slow reproductive rates.[52] Unlike most other fishes, which release huge batches of eggs and count on some of them being fertilized by sperm spewed nearby—and then having a small percentage of these fertilized eggs escape being consumed by predators—sharks reproduce by internal fertilization.[53]

When a female shark is ready to ovulate, she releases a chemical attractant that draws males. Because sharks generally lead solitary lives, body language won't do; but sharks have such an acute sense of smell that the male can notice that sexy scent across a crowded reef. Shark sex is rough. The male's external reproductive organ is called a clasper. He has a pair of these located along his underside, between his pelvic fins. Despite the name, he doesn't use his claspers to hold his intended in place; instead, he does this by biting one of her pectoral fins violently, sometimes leaving a lasting scar. Twisting her to one side, he inserts one of his claspers into her cloaca, a channel that serves excretory as well as reproductive functions. His sperm travels down a groove in the clasper, which is made of stiff cartilage, and up her fallopian tubes, fertilizing the ripe eggs. After he releases her fin, she swims off on her own to gestate.[54]

Throughout the world's oceans, shark reproduction varies from species to species. Some sharks lay eggs, each in its individual package with a yolk for nourishment, and anchor these in protected crevices in rocks or corals. When the baby shark is ready to enter the world, it exits via a slit in the side of the case. Beachcombers sometimes find these egg cases, given the fanciful name "mermaid's purses." Other species of sharks extrude their eggs in collective spirals. But all common Caribbean sharks carry their eggs internally until they are ready to hatch.[55] Hammerhead moms provide additional nourishment via an umbilical cord.[56]

Depending on the species, shark gestation ranges from 5 months to 24 (for the spiny dogfish shark, *Squalus acanthias*, the longest pregnancy of all vertebrates, on land or sea).[57] For sharks common on the Mesoamerican Barrier Reef, pregnancy lasts from 5 months for the nurse shark to as much as 16 for the tiger.[58]

When she senses that her long pregnancy is nearing its end, the mother shark heads for someplace warm and shallow, with abundant fish eggs, small crustaceans, and other petite prey suitable for newborn "pups," as baby sharks are called. For bull sharks that means brackish or fresh water, perhaps because bays and river mouths are more protected.[59] Bull sharks give birth to only 1 to 3 pups; hammerheads, 12 to 15; nurse sharks, 20 to 30. Tiger shark litters start out with 13 to 16; but in the case of sand tiger sharks, the first two pups to hatch devour their laggard siblings.[60]

Choosing a suitable nursery marks the extent of the female's maternal duties. Sharks are not mammals. They don't nurse their young, protect them from predators or teach them how to find their own food. On the other hand, newborn sharks don't need their moms' help. They emerge well-developed and able to fend for themselves. Except for their disproportionately large eyes, shark pups look and function like miniature adults.[61]

The most literally breathtaking shark in the Caribbean is the whale shark (*Rhincodon typus*), the world's largest non-mammal. With solid white bellies and lines of white spots decorating their gray or brown tops and sides, they cruise at depths of 20 to 30 feet over the deeper sections of the Mesoamerican Barrier Reef, sticking to the blue water. Coral heads and canyons are too shallow to accommodate their bulk. Each animal's pattern of spots is distinctive to that individual, like a human's fingerprint.[62]

Imagine swimming with a fish bigger than a bus. On average, adult whale sharks grow to 40 feet long, although individuals can reach 60[63]— half the length of a blue whale, but still twice the size of a spacious living room. In weight, whale sharks average more than 20 tons.[64]

Whale sharks and whales can get that big because they live in the sea, which supports them in two ways—the salt water that lends them buoyancy[65] and the plankton that nourishes them. In search of plankton, whale sharks travel huge distances and dive as deep as 3,000 feet.[66] Whale sharks feed constantly, swimming with their mouths open and using black sieve-like structures behind their mouths to filter krill, fish eggs, crab larvae, and whatever small fishes and crustaceans either wash through or get sucked in. The shark then expels the excess water through its gills.[67]

Whale sharks live about 70 years, spending those seven decades helping preserve the balance of life on the world's tropical coral reefs. Drawn by the spawning of cubera, dog, and mutton snappers, these marine mammoths gather at the Gladden Spit and Silk Cayes Marine Reserve, 26 miles off the southern coast of Belize, from March through June in the days around the full moon.[68] Dolphins and bull sharks join in the feast. The snappers gather at the full moon so that their fertilized eggs can float on the high spring tide out to the relative safety of the sea.[69]

As intimidating as the animals' size makes them, the only danger whale sharks present to humans is the risk of being hit by a massive tail. Dubbed the "gentle giants" of the sea, they greet divers and snorkelers with apparent curiosity and have even been reported swimming alongside boats, as if asking to be petted. (Responsible captains and dive guides don't allow touching whale sharks, for fear of injury on either side.) There's a lot that scientists still don't know about whale sharks, such as how and where they mate and how far they travel.

Even once we answer those questions, at least one mystery will remain: What do whale sharks make of *us*?

III

WHAT WORKS

Joining Forces to Save the Reef

15

~~~

INVASION OF THE LIONFISH

It was happy hour at Hamanasi, an upscale eco-resort on the Belize coast about 120 miles south of Belize City. The bar was large and airy, with couches and chairs upholstered in tropical prints and arranged in conversation groups. A bookcase stocked with board games and jigsaw puzzles sat against one wall, and two sets of French doors led out to the verandah and the lighted pool beyond. Behind a curving bar crafted of sustainable hardwood, the best mixologist in this part of the country was filling a silver shaker with the ingredients of the drink of the day, a concoction involving coconut rum, vodka, fresh pineapple juice, and blue curaçao.

Although the place was usually busy, this evening it was packed. Free appetizers provided the lure, and the head cook had laid them out on a table brought in for the purpose. Neatly lettered signs identified each snack—lionfish fingers, lionfish cakes, lionfish balls, lionfish ceviche. Various dipping sauces—buffalo-wing style, barbecue, jerk, mustard, sweet-and-sour—rounded out the presentation. The intention was clear: Get the bar's patrons to try the mild, slightly sweet-tasting meat of this fish, an invasive species with a stunning rate of reproduction and no natural enemies in the Caribbean. Once guests had sampled lionfish, they would be more inclined to order it as fish tacos at lunch or as the catch-of-the-day at dinner.

Lionfishes (red lionfish: *Pterois volitans* and its cousin the common lionfish or devil firefish: *Pterois miles*) are a prime example of how human thoughtlessness can disrupt a fragile ecosystem and how human ingenuity and cooperation may be able to control—or at least reduce—the damage. Although they are native to the Indo-Pacific, specifically to the tropical waters surrounding the Philippines, Indonesia, and the islands of the South Pacific,[1] in the early 1990s lionfishes began appearing in South Florida. The first confirmed sighting, in 1985, was treated as a curiosity, rather than a threat,[2] but around 2000 the population exploded and by 2010 lionfishes had settled in throughout the Caribbean and the warm waters of the Western Atlantic, from North Carolina to Brazil. By 2002 they had been sighted as far north as Long Island.[3] No one knows definitively how these invasive species were introduced to an environment half a world away from their home.

One guess marine scientists can offer is that the fishes were released intentionally from home aquariums by owners who didn't recognize their potential for destruction. With distinctive maroon or brown stripes set off against white, antennae jutting like horns from their foreheads, and tapered pectoral fins rippling like feathers, lionfishes are among the most exotically beautiful denizens of the sea. Small wonder that they became prized in the aquarium trade. But adults average 12 inches in length and can grow to 18, much too big for the typical saltwater tank in a living room. And when a family moves from Florida to Chicago, the aquarium doesn't travel full.

Some believe that lionfishes escaped from Florida pet shops that were destroyed by Hurricane Andrew in 1992, despite the earlier sightings off the Florida coast. Or perhaps juvenile lionfishes were flushed from the bilges of vessels plying the commercial shipping lanes between the Philippines and the Atlantic. This is how more than 3,000 species of aquatic life enter the world's oceans daily.[4] These hypotheses are not mutually exclusive.

Lionfishes probably developed their showy coloration to ward off predators, which they do fairly effectively, even in their native waters, where local groupers have developed a taste for them.[5] Although their life span in the wild is 15 years, one lionfish survived to the ripe age of 30 in an aquarium.[6] They adapt to all sorts of tropical environments, including coral reefs, seagrass beds, mangroves, and shipwrecks.

These canny invaders kill and eat nearly 170 marine species[7] and have stomachs that can expand thirtyfold.[8] Lionfishes destroy species valuable commercially and for sportfishing in two ways: they eat juvenile groupers, snappers, etc., and they compete—very effectively—with the survivors for food.[9]

When it comes to water pressure, lionfishes are amazingly adaptive, thriving from shallow reefs to profound depths. Researchers piloting submersibles have found them as deep as 1,000 feet.[10] And they are remarkably fecund. A female lionfish can spawn every four days and lay a whopping 30,000 eggs at a time.[11] Possessing these advantages and lacking natural enemies in their new Western Hemisphere home, lionfishes quickly became a threat to their adopted ecosystem.

By 2009 they had established themselves along the Mesoamerican Barrier Reef, where they competed for prey with snappers, groupers, and other native species, whose numbers were already declining due to overfishing. Swimming slowly by undulating their dorsal and anal fins, lionfishes use their long pectoral fins to trap their prey, largely small crustaceans and fishes, including those juvenile snappers and groupers mentioned previously.[12]

In nature guide books and fish identification charts, lionfishes are listed with such other dangerous marine life as scorpionfishes, sea wasps, and fire corals. And well they should be. To defend themselves against predators, lionfishes carry a powerful punch of venom in the ends of their dorsal fins. Each lionfish has 18 spines that deliver a nasty cocktail composed of a protein, a neuromuscular toxin, and the neurotransmitter choline. In humans, this venom causes sweating, extreme pain, respiratory distress, and even paralysis. Although it is only directly fatal to people with allergies to its components, pain from a sting could cause potentially lethal responses, such as bolting to the surface from a depth sufficient to cause decompression sickness ("the bends") or a pulmonary embolism.[13]

But although lionfishes are venomous, they are not poisonous.[14] To understand the difference, think about a rattlesnake. The snake's fangs pack a dangerous dose of venom, but its flesh is safe to eat and is occasionally served, primarily as a novelty, at restaurants in the Southwest, where customers compare its taste to chicken.

Claiming to offer the world's first invasive species menu, Chef Bun Lai used lionfish to propel both his New Haven, Connecticut, restaurant, Miya's Sushi, and its Miami South Beach spinoff, Prey, to international fame.[15] For his Kiribati Sashimi, Lai sliced raw lionfish thin and dressed it with a squeeze of lime, seven different varieties of crushed peppers, seaweed flakes, toasted sesame seeds, and sea salt from Kiribati, a Pacific island threatened by inundation due to climate change. For $20, diners could enjoy this environmentally friendly appetizer before proceeding to "La Soupe des Means Greenies," a miso soup with a base concocted from dried, pulverized carcasses of the green crabs that invaded U.S. waters during the nineteenth century; then moving on to some Asian sea squirt, considered a delicacy in South Korea but now taking over blue mussel territory from Maine to New Jersey, and perhaps adding an entrée of bow-shot mute swan, a bird that looks graceful but damages marshes. After rubbing the swan meat in oil, grated ginger, and jerk seasoning, Lai chopped it, mixed it with shallots and rosemary and served it rolled in a leaf of kudzu, the invasive vine native to Japan that has smothered forests throughout the South. For a beverage, Lai offered a lemonade substitute created from the knotweed that is crowding out native species in 39 states. Lai called the drink's taste "not unlike a Granny Smith apple."[16]

High in omega-3 fatty acids,[17] lionfish meat is a healthy, as well as eco-friendly, menu choice, and restaurants that serve it report a high customer satisfaction rate.[18] Lionfish is also lower in mercury and heavy metals than snapper, grouper, mahi-mahi, and other menu staples.[19] The sustainable seafood guide published by the Monterey Bay Aquarium and including fishes in the Caribbean gave "Best Choice" ratings to both red lionfish and devil firefish.[20]

Nowadays, commercial fishing accounts for more deaths of lionfish than any other cause, which means that efforts to persuade people to eat them are working, and communities along the Mesoamerican Barrier Reef are recognizing the economic benefit of harvesting them.[21] Human beings can be very effective predators.

Although London-based nonprofit environmental organization Blue Ventures has a global scope spanning a variety of programs, the focus of its efforts in Belize is lionfish control, including the promotion of the aquatic invaders as menu items. According to Jennifer Chapman, Blue Ventures's Belize country coordinator, the meat of the lionfish presents no risk to humans except in areas where the nerve toxin ciguatera can be present in any species of fish.[22] Quintana Roo, the Mexican state bordering the Mesoamerican Barrier Reef, reported 163 cases of ciguatera between 1984 and 2013; but these all resulted from eating snapper or barracuda, not lionfish.[23]

Beginning in 2003 with a project in southwest Madagascar, Blue Ventures has initiated programs with traditional fishing communities across the coastal tropics worldwide. Its goal is to produce both environmental and social benefits.[24] "We work in areas where a healthy ocean is vitally important to people and culture and look for solutions that produce maximum benefit for the coastal communities," Chapman explained.[25]

All projects have involved collaborations with the communities themselves and with local government, recognizing the effectiveness of community-based conservation and the ethical basis for taking a community-led, rather than a top-down, approach.[26]

In addition to a campaign to encourage Belizeans to eat lionfish, a species new to their diet, in 2013 Chapman's group set up a program for local women to make lionfish jewelry. For the most part, they utilize the tails, which are safe to handle. When they do use the spines, they bake them first, rendering them harmless. Heat breaks down lionfish venom, which is why the application of hot water is effective first aid for a sting.[27]

"Jewelry-making is a really fabulous way to increase catch values of the fish and to give economic benefit to women," Chapman said. The program's success has spawned another challenge: Because of the existing social structure, gender roles in Belize are very traditional, mostly dependent on men as the breadwinners. "But women also want to have their own incomes, so they make the jewelry," she explained.[28]

In the Indo-Pacific, predators kept lionfishes under control by eating them. Relocated to the Caribbean, where they have no natural enemies, lionfishes have been free to flourish unchecked. Introducing foreign predators would complicate the existing problem, but what about developing a taste for lionfish, not just among humans, but among local marine species as well? One promising lionfish-control strategy being tested along the Mesoamerican Barrier Reef is training moray eels and groupers to recognize the invaders as food by feeding them freshly speared lionfishes.[29] Caught on video, the groupers swallow the lionfishes from the back, while the eels shake and spin the prey, consuming it from the belly up, so that the venomous spines slide harmlessly upside down into the morays' tooth-studded maws. Apparently, the strategy is working. Moray eels and large groupers now also hunt lionfishes on their own; remnants of the invaders have been found in their stomachs.[30]

Dive operators along the Mesoamerican Barrier Reef have jumped enthusiastically into lionfish control. Guides often carry sharp poles with which to impale members of the species that they encounter while leading groups of divers among the corals. Although spearfishing is illegal in the marine parks that protect the waters around the Honduran Bay Island of Roatán, enforcement authorities make an exception for lionfish. According to Ian Drysdale, Honduras in-country director for the Healthy Reefs Initiative (HRI), the parks sell a distinctively bright yellow version of the Hawaiian Sling to anyone who completes the two hours of training required for a special lionfish-spearing license.[31]

Hamanasi divemaster
Sam Norales spearing
lionfish, photo by Chris
Liles.

Some dive resorts sponsor periodic "lion hunts" to attract clients and involve them in eliminating these invaders. Such efforts can be stunningly effective, at least in the short term. In 2015, derbies in the Bahamas reduced lionfish populations by 70 percent in a 58-square-mile area.[32] During a one-day "hunt" off Jacksonville Beach, Florida, two divers removed 815 lionfishes.[33]

For the June 2018 issue of *Smithsonian*, Jeff MacGregor covered Pensacola's Lionfish World Championship, one of Florida's dozens of competitive lionfish hunts. In 2016 it brought in 8,000 of the captivating invaders.[34] In 2019 the number jumped to more than 19,000.[35] The annual event included a lionfish food festival, in which some of the region's celebrity chefs participated and people lined up for everything from fried filets to lionfish-stuffed jalapeño poppers.[36] The slogan for the food fest was "Eat 'Em to Beat 'Em."[37]

Unfortunately, after lionfishes are thinned out, they tend to recolonize. Despite lion hunts and promotion of lionfish dishes, populations of this wily invader continue to swell in the Caribbean and the coastal waters of the southern and eastern United States,[38] prompting some marine biologists to liken removal efforts to pulling weeds in the yard. They always come back.[39] Eradicating a species that has become so well-established in its new home may be impossible, but controlling it may be feasible—if only we knew more about it. For example, knowing how far the fishes migrate would help pinpoint the origins of re-colonizers.[40]

The University of Quintana Roo, the Mexican state that borders the Mesoamerican Barrier Reef, is conducting studies of lionfish behavior.[41] So is the Reef Environmental Education Foundation (REEF) in collaboration with the National Oceanic and Atmospheric Administration (NOAA). Since 2008 they have been catching lionfishes in nets in the Bahamas, tagging them with plastic strips, and releasing them live back into the sea, hoping to monitor where they go.[42]

One challenge is that they probably go very deep. Lionfishes seem to be able to tolerate a vast range of pressure, from within a few yards of the surface to 1,000 feet down. When they plunge below 120 feet, recreational divers and dive guides can't spot them, so the tagged lionfishes don't get reported. One answer to controlling lionfishes on the deep reef may be traps. At the Pensacola tournament, Stephen Gittings, chief scientist for NOAA's National Marine Sanctuary System, showed Jeff MacGregor several models of a bell-shaped device designed to trap lionfishes hundreds of feet below the surface.[43]

Hoping to be able to monitor this invader wherever it goes 24/7, scientists are shifting to electronic tagging, using an acoustic device the size of an AAA battery. Inserting this device requires five or six minutes of surgery under anesthetic. Still, this demanding monitoring will be worth the trouble if it yields insights that will help keep one of the world's most effective, efficient invaders from crowding out the native species on which the health of the Caribbean's coral reefs depend.[44]

Unlike lionfishes, some foreign species never take hold in their new environment—the conditions simply aren't favorable. Others find the conditions, like Goldilocks's porridge, "just right" for settling in. Including animals and plants, on land as well as in the sea, scientists have identified 552 invasive species in the Caribbean region,[45] but apart from lionfishes the only one that comes close to presenting a current threat to the

Mesoamerican Barrier Reef is the orange cup coral (*Tubastraea coccinea*). With prominent tubes of pastel yellow or orange, it forms clusters that look quite innocent by day. After dark its bright yellow polyps send out tentacles to feed. Although it is a hard coral, it is not a reef-building coral; it competes, very effectively, for underwater real estate on which reef-building corals could otherwise form colonies.[46]

Like lionfishes, this invasive species is native to the tropical Indo-Pacific. First observed in Puerto Rico and Curaçao in 1943, orange cup coral was spotted in 1948 attached to the hulls of ships coming from the Philippines. Cup coral larvae can float for up to 14 days, so the species has spread easily on the Caribbean's currents. It reproduces vigorously, becoming sexually mature at just 1.5 years. Even more alarming, orange cup coral has an effective system of chemical defenses that prevent other corals from settling nearby and keeps reef-cleaning grazers like parrotfishes from eating it.[47]

So far, the invasive species that have presented the greatest potential threats to the Mesoamerican Barrier Reef have come from the tropical waters of the Indo-Pacific, from which they have hitched rides on planes (for example, as aquarium fish) and on hulls and in bilges of oceangoing vessels. The rise in global trade can only increase this trend. Since 1914, the chief route from the Pacific to the Caribbean and the rest of the Atlantic basin has been the Panama Canal. On June 26, 2016, an expansion opened, almost tripling the tonnage that could pass through this waterway daily. A third lane, locks widened to 70 feet and the addition of the Pacific Access Channel, a shortcut on the Pacific side, mean both more ships and bigger ones—and thus more opportunities for invasive species to travel from ocean to ocean. [48]

Checking the hulls and bilges of each of these vessels would be unfeasible, so these stowaways will keep coming. This particular result of globalization provides all the more reason to ramp up our efforts to detect and control invasive populations once they arrive. After all, we have the tools, from high-tech monitoring strategies to patronizing restaurants that serve lionfish tacos.

16

THE VIEW FROM CARRIE BOW CAY

The Importance of Monitoring

In February 2016, my dive buddy Terry McNearney and I spent nine days at Hamanasi, the Belize eco-resort where the cook concocted the lionfish appetizers described in chapter 15. On several of those days, the wind blew onshore with enough force that the dive operator decided not to take the boat out to the barrier reef, but as I described in chapter 8, on the days we did dive we experienced a wide range of environmental conditions. In some places, the Mesoamerican Barrier Reef looked healthy; in others, it appeared to be in serious trouble.

Beneath the algal gunk on Carrie Bow Cay Wall, the hard corals that hadn't been completely smothered seemed healthy, free of bleaching. Various species of reef-building corals layered on top of each other, with sponges and soft corals, sea fans and sea whips on top of those, reaching for the sunlight like the trees and vines in the nearby jungle did. The coral heads were cut with ledges, under which we saw five spiny lobsters, their long, pointy antennae sticking out and tapping around.

A spotted eagle ray with a wingspan of about nine feet glided eerily out of the murk, and despite the turbidity, we were able to spot a sizable Nassau grouper, plenty of parrotfishes (both the neon-hued striped parrotfish and the hipper-looking stoplight variety—black with red fins), queen angelfishes, French angelfishes, and big schools of blue chromis and other small blue fishes. I saw lots of juveniles inside vase sponges, including a small crab in a translucent white one and in a bright lavender one, a juvenile spotted drum, its delicate black and white striped dorsal fin waving like a ribbon. The sponges lent a bright contrast to the overall monochromatic pallet, where even the abundant gorgonians were mostly black or brown.

After the required one-hour surface interval to allow ourselves to off-gas some of the excess nitrogen we'd absorbed underwater, Terry and I dived the wall off South Water Caye, a few hundred yards to the north. (As mentioned in chapter 2, the names of the islets that dot the reef are sometimes spelled with an *e*, sometimes without.) Here the visibility was better and the reef more colorful, with plentiful pink and lavender accents. The structure was spur-and-groove, narrow patches of sand sloping between walls of coral. On one sandy stretch, 10 conchs in various stages of maturity waddled along.

Safety guidelines recommend against scuba diving within 23 hours of a flight, so I spent the day before my departure from Hamanasi on a snorkeling excursion. Our boat's first stop was the lagoon inside the fringe reef that surrounds Carrie Bow Cay on three sides like an elongated letter C, coming much closer to shore on the west side than on the north and south.

The water was gin-clear, with visibility of at least 80 feet. A startling variety of hard corals rose to the surface. Some of them were chimeras, with two different species occupying one coral head. My guide, Medel, a local whose neatly trimmed goatee gave him a vaguely hipsterish look, skimmed along the tops, sometimes with no more than 15 inches of clearance. With his sharp eyes and obvious familiarity with the site, he used a lionfish-hunting pole, which resembled a long broomstick with a two-prong fork at one end, to point out squirrelfishes, lobsters, and other denizens of the coral heads. A small black eel darted out from under a ledge. One of the advantages of snorkeling, as compared to scuba diving, was that after he'd shown me something, Medel could pop to the surface and tell me what it was.

Climbing back aboard the boat, we moved on to a site called Third Cut, so named because it was the third of five natural channels that cut through this stretch of reef. A couple of graceful yellow butterflyfishes swam by, sporting near their tails the large white-rimmed black dots that potential predators mistake for eyes, concluding that they've encountered a fish too large to take on. We saw three sea urchins—two short-spined and one long-spined. Looking like a chamois cloth with a tail attached, a yellow stingray rippled past beneath us. Medel showed me a sand filefish, only a few inches long and almost translucent, burrowing under a layer of broken branch coral. Pushing his snorkel to the side, he explained that the rubble adhered into a kind of concrete, forming a handy roof for the filefish's burrow.

Was the Mesoamerican Barrier Reef the vibrant ecosystem I had observed snorkeling or the murky underwater dystopia I had witnessed while scuba diving on the wall no more than a mile away? Curious and confused, I returned to this section of Belize six months later. This time I went to Carrie Bow Cay itself to spend a week observing the scientists at the Smithsonian Institution's field station. Researchers come to Carrie Bow in teams. First, they have to submit a proposal to the Smithsonian. If the Smithsonian approves their project, the scientists pay $100 apiece per day for rustic accommodations, hearty Belizean fare, unlimited scuba tank refills, and the use of wet and dry labs equipped with state-of-the-art microscopes and other scientific gear. In the first 45 years of the program, these visiting scientists had conducted more than 900 scientific dives and published more than 930 articles in scientific journals.

As one of hundreds of little islands that had emerged when sea levels dropped during the last Ice Age, Carrie Bow lies directly atop the Mesoamerican Barrier Reef, offering a ready perspective on what is happening, for better but mostly for worse, to this imperiled international treasure.

The Smithsonian had graciously complied with my own request to spend a week on Carrie Bow Cay, so that I could describe some of the work taking place there and the people conducting it. Zach Folz, station manager and dive officer for the Smithsonian's Caribbean Coral Reef Ecosystem (CCRE) Program, had secured the agreement of Karen Koltes, who held a doctorate in biological oceanography and was the leader of the team of four scientists who would be on the island for the first two weeks of September 2016. I would arrive at the start of the second week, to give them a chance beforehand to set up the monitoring of the reef that Koltes, with her husband John Tschirky, an expert in seagrass communities, had conducted almost annually for the past two decades.

Bouncing across the lagoon in a panga, the steep-prowed open boat favored by local commercial fishermen, I watched the pastel frame buildings of Dangriga recede and the staggered indigo ridges of the Maya Mountains fill the view to the southwest. To the northwest, the coast flattened into mangroves and jungle. After 20 minutes, little dabs of green appeared to the east, the low-lying mangrove islets that marked the crest of the Mesoamerican Barrier Reef. I still couldn't make out Carrie Bow Cay—it was too small. Only after patches of turquoise, signaling sand flats below, had begun to interrupt the dark aqua over the seagrass beds did I

The Smithsonian Institution's Carrie Bow Cay Field Station, photo by Mary Rice for the Smithsonian Institution.

see ahead the clump of coconut palms and the slender dock extending toward us. To its right was a one-story structure sporting a mural on its seaward wall—a huge manta ray, plus assorted smaller fishes and sea fans. Straight ahead about 10 yards from the foot of the dock, a white banner, nailed at its four corners to a two-story frame building, bore the familiar gold-on-blue Smithsonian sunburst and read: "Smithsonian Institution, Caribbean Coral Reefs Ecosystem Program, Carrie Bow Cay, Founded 1972 Rebuilt 1999."

As I climbed onto the dock, a large man with red hair and a red beard came striding toward me, looking like the embodiment of Hägar the Horrible from the Sunday comics. Sticking out his hand, he said, "Welcome to Carrie Bow Cay" and introduced himself as Craig Sherwood, the on-site volunteer station manager.

Within a few minutes, the boat pulled away, and Sherwood began my tour of the field station. Apparently I wasn't the first person to liken him to the comic strip Viking. He explained proudly that he came from Swedish and Finnish stock. Sherwood's assignment was to keep the physical plant running—the rainwater collection system, the solar power, the three boats, the satellite communications, the compressor for the scuba tanks, the composting toilets.

Sherwood started with the main building, a two-story Z-shaped frame structure with verandahs around two sides, top and bottom. Like the other three buildings on the cay, it was painted light blue with white trim. Showing me around the downstairs, he pointed out a couple of bedrooms, two mechanical rooms, the kitchen, the open porch with the long table where everyone ate their meals, and the wet lab. Off to the side and a few steps down was the saltwater lab; three of its walls were lined with shelves, with stacks of five-gallon plastic buckets beneath them.

Leading me up an open stairway, Sherwood pointed out the dry lab, where Karen Koltes and her team had spread out their laptops amid the stereo and compound microscopes and other sophisticated equipment. (They were currently out on a dive collecting data.) Across the hall was the library, which contained scientific publications, marine life identification guides, and assorted beach books, mainly dog-eared thrillers and mysteries. It also had a video setup and several shelves of DVDs.

Next on my tour was the building with the mural. On one side was the dive shed, with the scuba tanks and the compressor that Sherwood used to fill them. On the other side were two generous-sized shower stalls. With both the temperature and the humidity in the 90s, Sherwood had turned off the hot water.

The prettiest structure on the island was the Honeymoon Cottage, a raised gingerbready affair with its own composting toilet and sink. Koltes and Tschirky were staying there. I would be bunking in the middle unit of a three-room cabin. My digs were 6 feet wide and 10 feet long, with a few wooden pegs and a three-inch shelf along one wall. An alcove about three feet square contained a little table and a wooden chair. The windows lacked screens, and there were no fans or electrical outlets; but a broad verandah ran along the seaward side of the building offering two rope hammocks and a fine view of the fringe reef and the narrow lagoon between it and the cay. Brown pelicans splashed into the shallow water. Looking like avian versions of Charlie Chaplin, ruddy turnstones scurried along the shore. Green geckos darted up the support posts.

Clanking and the buzz of conversation alerted me that Karen Koltes and her two assistants had returned from their afternoon dive. John Tschirky helped them unload their gear. A tall, slim man with collar-length brown hair that was going gray at the temples, Tschirky looked like he belonged on a sailboat. He had been living on a boat off the north shore of the Dominican Republic when he had met Koltes, who at the time managed a research station on Grand Turk. They had been attending a conference for a project on reef health. Now Koltes was coral reef program manager on the Policy Team for Insular Affairs at the U.S. Department of the Interior, and Tschirky ran the American Bird Conservancy. Both were using vacation time for their two weeks on Carrie Bow Cay.

Since suffering a double pneumothorax underwater a few years earlier, Tschirky had retired from scuba diving. For some reason that his doctors never figured out, air had filled his chest, leaving him temporarily with only a quarter of a lung. Although he had recovered the use of both lungs on land, he couldn't risk subjecting his body to several atmospheres of pressure and the challenges of breathing compressed air. However, he continued to monitor the seagrass beds by snorkeling and by conducting analyses in the lab.

Like Tschirky, I wouldn't be diving with the researchers. For me the reason was that I wasn't qualified as a Scientific Diver, a certification that requires the successful completion of a rigorous course approved by the American Academy of Underwater Sciences. (The one taught at Texas A&M University at Galveston was 100 hours long.) To console myself, I had booked five nights following my week at the field station at Isla Marisol, a dive resort on Glover's Reef, the remote Belize atoll mentioned in chapters 2 and 4.

The Koltes team's days were packed with conducting observations, collecting samples, and recording what they learned for future analysis. Not until almost sunset on my fourth evening on the cay did Koltes have time for me to ask her for her assessment of the health of the Mesoamerican Barrier Reef. Watching the Caribbean turn from turquoise to aqua as the sun set behind us, she and I settled into Adirondack chairs on the second floor verandah of the main building, sipping One Barrel, the dark, drinkable, and very cheap Belizean rum. Koltes's ash-blond curls had dried since her afternoon dive. A strip of Day-Glo green electrician's tape wrapped her right middle toe where it had been rubbed raw by her scuba booties.

Koltes told me that except for the three-year post–Hurricane Mitch period when the field station was being rebuilt, she and Tschirky had been monitoring this spot on the Mesoamerican Barrier Reef for 20 years.[1] They began the program in 1993 in response to rising concern about the viability of the Caribbean's coral reefs.

In the 1980s there had been problems with the reefs; one of the earliest documented was a bleaching event in Florida. As the introduction and chapter 1 pointed out, bleaching doesn't always kill corals directly, but it can leave them vulnerable to potentially lethal pathogens, such as the white band disease that jolted environmentalists and marine scientists, including Koltes and Tschirky, by killing staghorn corals around Florida and the Bahamas. Eventually, the white band disease spread west and south, destroying 95 percent of the staghorns in the Caribbean. Adding to the decline was the overfishing of herbivorous species that kept the reef clear of damaging, overabundant macroalgae. Spiny sea urchins had also been doing the job. (As described in chapter 9, the urchins would later succumb to a mysterious plague themselves.)

"Things were looking pretty dire," Koltes told me as we sat gazing out at the Caribbean. "A bunch of scientists got together and said, 'How can we say something is deteriorating if we don't have measurements?'"[2]

So Koltes and Tschirky applied to the Smithsonian Institution for permission to use the Carrie Bow Cay Field Station to monitor life along the reef. The scientists who had come to the cay during the 1970s had concluded that in terms of animal and plant life, the conditions on the cay, which sat right on top of the barrier reef at 16°48'N and 88°05'W,[3] were similar to those elsewhere along it.[4] Carrie Bow Cay would make an ideal base for a longitudinal study of the tropical marine ecosystem and how it was changing.

To ensure that they would be measuring the same things in the same places every time, Koltes and Tschirky laid out transects, the scientific term for lines connecting two points on the earth's surface. Each of their 10 transects consisted of 10 meters of common window sash chain with half-inch links, anchored at the ends with steel posts. Coming back to the same spots year after year, a team of scientists could monitor what was growing on, under, and around each link. To keep track of positions along the chain, Koltes and Tschirky marked every tenth link with colored strips of plastic.[5] Odd tens (10, 30, 70, and 90) were white; even tens, blue. Every fiftieth link was orange; every hundredth, black. Link 500 bore a piece of pink flagging tape.

"All this helps me 'map' where I am, so I don't get lost when I take my eyes off the chain to write down the data," Koltes explained.

During her monitoring dives, Koltes recorded what she found on Xerox waterproof paper called "NeverTear," attaching the sheets to a Plexiglas clipboard with rubber bands. She used 10 sheets, one for each transect. For ease of recording, she printed rows and columns ahead of time. The link number went in the first column; the species, each with its own four-letter code, in the second. She wrote underwater with soft-leaded Japanese "School Girl" brand pencils. The ones she used during the week I was on Carrie Bow Cay had images of Hello Kitty decorating the shafts. Once Koltes was back on land, she recorded the data on her computer, then rinsed and dried the paper and kept it for her permanent record.[6]

Her assistants focused their transect surveys on soft corals—the gorgonians, shaped like fans or feathers, that wave in the current, filtering food that passes by. Koltes printed out the data from the previous year on the same underwater sheet with the species name, location, height and width, and any comments. That made locating the soft coral colonies much less confusing. Koltes marked the ones that needed follow-up in yellow and the ones whose identification had been confirmed by microscopic examination in the laboratory in green. That way, the scientists didn't need to take any more samples than necessary.[7] By 2000 all the affiliated researchers studying the reefs of the Caribbean had adopted the same method for data collection, making comparisons clear and analysis more powerful.

"There's a huge diversity of research that goes on here at Carrie Bow Cay," Koltes said. She wanted to be sure I understood that what she and Tschirky and their team were doing was not technically research. "My group is monitoring, not testing hypotheses," she explained. "We're tracking trends."[8]

Many marine scientists, including those who utilize the field station, devote their time to generating and testing hypotheses. But the results of monitoring often inspire these hypotheses, and both active research and monitoring can result in articles in peer review journals and thus in tenure and promotion for people in academia.

Pausing, Koltes took a more confident tone. "I guess what we're doing is natural history," she said. "We need more natural history."[9]

When Koltes and Tschirky started out, they made their first transect on the original one done by the Smithsonian team in 1972. That gave them a baseline—not the baseline of the pristine reefs encountered by Spanish sailors in the sixteenth century, but a baseline nonetheless. Koltes and Tschirky had replaced the transect only once, after it had been destroyed in a 1997 fire that had engulfed the field station.[10]

As they laid their second transect, this one south of the cay, they noticed a lot of dead staghorn coral. Staghorn coral colonies form stalks about the diameter of a human thumb that branch off from each other and end in white tips. The dense clumps look like piles of antlers. When I first started diving, in the early 1980s, staghorn was so abundant on Caribbean reefs that scuba divers and even snorkelers had to be mindful not to run into it and incur painful pokes. Now it was on the Endangered Species List, like its cousin elkhorn coral, which grows similarly but ends in scalloped plates. Both are vulnerable because they're top-heavy, like trees rather than boulders or pinnacles. Bleaching can weaken the elkhorn and staghorn "trunks" that attach them to the reef, and the next sizable storm surge can topple them.

From the surface, the Mesoamerican Barrier Reef looked healthy, almost pristine. Granted, replacing the mangroves with sand and coconut palms had caused Carrie Bow Cay to shrink from the two acres it had covered in the 1940s to less than an acre today. But the environment was still healthy enough for a mother loggerhead turtle to emerge from the sea early my second morning and wander around in search of a place to nest, as I described in chapter 4. Later that same day, at a break in the fringe reef about 20 yards from where I sat, the biggest osprey I'd ever seen devoured a silvery fish about 18 inches long. Using its left talons to balance on the coral, the raptor held its limp meal in its right, leisurely tearing off chunks of flesh with its scimitar-shaped beak. In the shallow lagoon between the reef and my own perch, a four-foot-long nurse shark cruised, as if waiting for the osprey to drop a morsel. Keeping their distance from the shark, pelicans dive-bombed a shimmering school of small fishes in water as clear as air.

As part of their annual monitoring, Koltes and Tschirky were keeping records of water clarity. They used two methods. One was a network of inexpensive waterproof light meters.[11] The other involved lowering a Frisbee-sized white disk, called a Secchi Disk, from the surface. Points along the rope were marked. When the disk vanished from view, the team member made note of the depth.[12] The method was similar to what Mark Twain had used to take soundings from a paddleboat along the Mississippi.

Hard corals are picky, and not just about temperature. They like clear water low in excess nutrients such as those that result from sewage, industrial activity, and agricultural runoff. A healthy reef requires clear water. Compared to reducing the greenhouse gases implicated in global warming, making the case for improving water clarity was relatively simple. It didn't take a degree in oceanography to understand that coastal development and population growth had resulted in more runoff and seepage from the land; that fueled by the glut of nutrients phytoplankton had gotten out of hand, reducing visibility along the Mesoamerican Barrier Reef significantly; and that building a new, more efficient water treatment plant and seeing that all homes and businesses were hooked up to it would help address part of the problem. Politicians and community leaders could understand this. So could the representatives of agencies, nonprofits, and philanthropies capable of funding the project. Cleaning up the water wouldn't be cheap, but it was a realistic goal, and it would improve the health of the local people, as well as the health of the reef.[13]

Simple silt is also harmful for corals. Silt comes from erosion, which is accelerated by coastal development, the destruction of coastal mangroves, and the rising intensity of tropical storms due to climate change. Seven named storms, six[14] of them full-blown hurricanes, hit the Mesoamerican Barrier Reef during the 2020 Atlantic hurricane season, which extends from June 1 to November 30, flooding the adjacent mainland with torrential rains that flushed pollutants and silt into the lagoons and onto the corals. The eye of compact but fierce Hurricane Delta passed south of Cozumel on October 7, just four days after Tropical Storm Gamma hit the same area.[15] Slow-moving and exceptionally wet Hurricane Zeta, which made landfall October 26, dumped even more rain on the Yucatán.[16] In early November unusually late and nasty Eta intensified from a tropical storm to a Category 4 hurricane so quickly that meteorologists were stunned.[17] On November 16 another Category 4, Hurricane Iota, bashed into Central America.[18] The eye made landfall in northeastern Nicaragua just 15 miles from where Eta had hit two weeks earlier. Bands of torrential rain flooded the rugged, already saturated landscape as far north as Mexico, flushing silt and pollutants from the mountains, farmland, and cities onto the reef.[19]

In that part of the Mesoamerican Barrier Reef, sediment from runoff delivers a double whammy, first as it floods down rivers and streams into the lagoons and onto the corals and then in a current-driven gyre that circulates silt and pollution back several months later.[20]

Hurricanes are increasing in both number and intensity.[21] The wind-fueled surf pounds tropical reefs, battering hard coral colonies, breaking sponges and soft corals, killing marine animals, and depriving those that survive of their habitats. Ruptured fuel tanks spew oil into the water. Debris from boats, houses, and docks smashes into the coral. Torrential downpours deluge wastewater treatment plants and septic tanks, spilling raw sewage. People on the islands and along the coast lose their homes and commercial buildings. Even if their boats survive, fishers face weeks without catches.[22]

But the most serious and lasting damage may come from the runoff. Silt kills.

Suspended, silt keeps sunlight from getting through to the corals. Once it settles, it smothers the polyps.

"The siltation events are getting more frequent," John Tschirky told me as I watched him wash a clump of seagrass in the saltwater lab. "A lot of the silt doesn't leave the system. It just settles. The water is never as clear as it was, because a smaller event, like a squall, can stir the silt up from the lagoon bottom and the seagrass."[23]

In their two decades of monitoring water clarity at Carrie Bow Cay, he and Karen Koltes had documented a constant downward trend, bad news for reef-building corals.[24] Protecting seagrass meadows and encouraging their growth would, in turn, help improve conditions for the entire Mesoamerican Barrier Reef.

Tschirky showed me the little holes the research team had made in the blades. Like human fingernails, seagrasses grow from the base, not the tip. By marking the spot where the white met the green, then measuring how far the hole moved up the blade in the course of a couple of weeks, Tschirky could determine how much the grass had grown. After rinsing the grass thoroughly and taking those measurements, he put the clumps in a drying oven, which looked a bit like a ceramics kiln but operated at a much lower temperature. When the grass had dried completely, its resulting weight would constitute its biomass—the living material less the water. Increase in biomass would indicate improving health of the seagrass beds that provided shelter to the juvenile crustaceans and fishes that would populate the reef as adults.

In 2016 the Mesoamerican Barrier Reef was still feeling the effects of Hurricane Mitch. When the monster storm hit in 1998, the accompanying rains caused massive runoff in Honduras and Guatemala. Carried north by the Caribbean Current, the silt was continuing to slosh around the reef almost two decades later. Overall, what Karen Koltes and her team found was a decrease in water clarity, which was very bad news for corals.

"A lot of what we're doing is monitoring the recovery from this cata-strophic event," Koltes told me. "The shallow reefs have recovered nota-bly; but ten meters deep, they don't seem to have. It may be a combina-tion of depth, light, and water clarity."[25]

Not everything was looking worse. "I expected to see more corals bleaching as a result of this year's El Niño, but was surprised to see rela-tively little," Koltes noted.[26]

She and her team were collaborating with other groups of scientists on the monitoring. Changes in such factors as water temperature might be relatively small, but taken several times a year, year after year, at the same spots on the reef, they demonstrated trends.

"We have found some interesting things as a result of long-term moni-toring," Koltes said.[27] For example, soft corals, such as sea fans, were re-placing the dead hard corals. Whether that was good or bad depended on the perspective of the species and the animals that depended on it. De-spite the stresses it was confronting, the Mesoamerican Barrier Reef was anything but lifeless. "Compared with the rest of the Caribbean, Belize has some of the better reef conditions," she added. "They've got a handle on managing the reefs."[28]

For the researchers who assisted Koltes and Tschirky, the weeks on Carrie Bow Cay offered an opportunity to participate in work that had the potential to inform positive change. Both Joanna Walczak and the other scientist, who asked that I not use her name, were administrators for the Florida Department of Environmental Protection. Like Koltes and Tschirky, they were using vacation time to put in a demanding two weeks on Carrie Bow Cay. And they were loving it.

"I don't have to worry about anything here," Walczak explained. "I get to be a scientist again."[29]

I spent much of my own week on Carrie Bow Cay reading publications that John Tschirky had gleaned for me from the field station's library. A thick, 9" × 12" volume contained academic papers written by scientists who had explored the barrier reef and seagrass beds around Carrie Bow Cay in the 1970s and early 1980s.[30] Here were detailed descriptions of everything from the geology of the reef to the distribution of species to the chemical defenses some algae use to discourage parrotfishes and sea urchins from grazing on them. Almost every paper held gems of informa-tion that I would be able to use in this book.

Those first teams of marine scientists who visited Carrie Bow Cay made some remarkable discoveries, in many cases documenting them with photographs. Their impressive work included identifying heretofore unknown species. But in most cases, the researchers didn't come back to find out what had happened on, and to, the Mesoamerican Barrier Reef. That would be left to Koltes and Tschirky and others willing to conduct the comparatively repetitious work of monitoring—the natural history.

Monitoring might not be sexy, but it is essential to identifying long-term trends. "It is very important that we continue supporting scientific monitoring," Mexican marine biologist Omar Vidal told me several years later. "It is very important for us scientists and for NGOs to get funds to continue monitoring how the reef and the communities are adapting to climate change."[31]

The morning of September 14, the four scientists on Carrie Bow Cay bustled around, logging their last findings, loading equipment and paper records into crates. I stripped my bed, packed my bag, paid my drinks tab for the week, said my thank-yous and goodbyes, and did my best to keep out of the way.

Around 3:00 the weekly supply boat from the mainland pulled up to the dock. In addition to food stuffs, toilet paper, and cases of Coke, Fanta Orange, and Belikin, the local beer, it bore a widowed river guide from Boise, who would be replacing Craig Sherwood as volunteer station manager. The boat taking me to the Isla Marisol dive resort on Glover's Reef was scheduled to pick me up half an hour later.

On his way to the dock, Sherwood gave me a hug and wished me good luck with my book, adding, "If this reef wasn't here, I would feel empty."[32]

"So would I," I responded.

"So should we all," I told myself, gazing first to the north and then to the south along the crest of one of Earth's most remarkable, and most fragile, wonders.

17

THE LOBSTERMEN'S DILEMMA

A Model Solution

In an open-sided building at the foot of one of the two docks in Punta Allen, also known as Javier Rojo Gómez, a little lobstering village on the Bahía de la Ascensión in the Mexican state of Quintana Roo, a meeting of the Vigia Chico Fishing Cooperative is about to start.

Punta Allen lies just 24 miles from Tulum, the ancient, spectacularly situated Mayan ruin that marks the southern extent of the 80-mile-long Riviera Maya, one of the fastest-growing tourist areas in the world. The Riviera Maya includes Cancún and Playa del Carmen, as well as Isla Mujeres and Cozumel just offshore and atop the Mesoamerican Barrier Reef. But Punta Allen is protected from that bustle, not just because it sits at the end of a two-hour dirt road but also because it is part of the Sian Ka'an (Mayan for "Place Where the Sky Is Born") Biosphere Reserve, established in 1986 and declared a UNESCO World Heritage Site in 1987.[1] The reserve's lagoons, wetlands, and reef are home to thousands of species, including tapirs, jaguars, and howler monkeys. Recreational fishing is permitted on some of the sand flats, but commercial fishing is prohibited, except for the lobstering rights granted to the Vigia Chico Fishing Cooperative and two others. Although the cooperatives hold 20-year licenses, they, as well as each of their members, must requalify for these annually.[2] They must honor the four-month closed season, from March 1 to June 30.[3] And they must use only the fishing gear and methods approved for the reserve—no scuba gear or hookah hoses attached to external air supplies. Lobstermen have to earn their catches the hard way, by free diving.[4]

As noted in chapter 11, pound for pound, the indigenous spiny lobster (*Panulirus argus*) is the most commercially valuable creature in the Caribbean.[5] It is arguably even more valuable ecologically, forming an essential link in the Mesoamerican Barrier Reef ecosystem, keeping some species in check, providing food for others.

Measuring more than a thousand square miles, the Sian Ka'an Biosphere Reserve incorporates two bays, each about 120 square miles—Bahía Espiritu Santo, with Punta Allen at its northern end, and Bahía de la Ascensión, the next bay south. The reserve also includes the jungle around the bays, the lagoons between the mainland and the Mesoamerican Barrier Reef, the cays dotting the reef, and the reef itself.[6] To the south and 22 miles offshore is a 310-square-mile atoll called Banco Chinchorro, another Marine Protected Area and UNESCO World Heritage Site, where only local fishing cooperatives are allowed to take lobsters.[7] Together, Sian Ka'an and Banco Chinchorro have six fishing cooperatives. The fishermen control the cooperatives, directly electing an administrative council consisting of a president, secretary, treasurer, and vigilance member. The council manages the cooperative's finances and business dealings, as well as its permits, concessions, and relationships with government authorities.[8] Decision making is bottom-up, with the fishers belonging to the cooperative deciding, collectively, what the rules should be and how they should be enforced and passing along what they decide to fisheries managers and park rangers and other authorities.[9]

Officially, Mexico's National Commission for Protected Areas (CONANP) and National Commission for Fisheries and Aquaculture (CONAPESCA) are charged with enforcing regulations in Marine Protected Areas,[10] but fishing communities like Punta Allen are often so remote and law enforcement resources stretched so thin that this isn't effective. Having the fishing cooperatives agree, in assembly, to incorporate the regulations into their own rules and having the members responsible to each other and to the group works much better.[11]

Mexico began establishing fishing cooperatives during the land reforms of the 1930s as a maritime equivalent of the *ejidos,* communally managed farms and ranches. Although the initial purpose was social and economic justice,[12] the six Mexican fishing cooperatives that share rights to protected portions of the Mesoamerican Barrier Reef have proven effective in gathering together the fishers, fishery managers, and other stakeholders to discuss and understand what they all have to gain by "sustainable" lobster harvesting,[13] in accordance with the World Commission on Environment and Development's definition of "sustainability" as "development that meets the needs of the present without compromising the ability of future generations to meet their own needs."[14]

Persuading the fishermen of Vigia Chico and the other five Sian Ka'an and Banco Chinchorro cooperatives to embrace sustainability took time. The little village of Punta Allen had become famous for its lobster harvest—not just the number of lobsters, but their size.[15] Mexican law forbids taking lobsters with carapace lengths under three[16] inches, but as this book goes to press the country's regulations impose no upper limit. Commercial fishers naturally target large adult lobsters, but these lobsters are the most valuable reproductively, as well as commercially, because size is related to fecundity. Larger females produce more eggs and larger males more sperm, and large females tend to spawn twice a season, compared to only once for small females; so removing the biggest individuals causes a disproportionate depletion of the population's reproductive potential.[17]

Overfishing has a spiraling effect. Because it causes a decline in the number of lobsters available, the ones that remain increase in value, giving fishermen an added incentive to go after them.[18] But these ever-scarcer lobsters also require more work to catch. Noticing that their stock was declining, the lobstermen of Sian Ka'an and Banco Chinchorro sought to balance short-term economic needs with long-term stability—for themselves, for their communities, for the lobster population, and for the environment.

Calling on experts in everything from marine ecology to marketing, the cooperatives developed a multilevel program for sustainability—not just for the lobster population, but for their communities as well. After studying their efforts for his doctoral dissertation from Curtin University in Perth, Australian marine biologist Kim Ley-Cooper noted that one factor that is making the sustainable lobster fishing program work in Mexico is that the people in the villages have felt protective of their sections of the Mesoamerican Barrier Reef since long before they were declared Marine Protected Areas. Ley-Cooper hoped that sustainable practices for fishing all commercial species would spread to other communities in the four countries that share this unique ecosystem.[19]

However, achieving a true commitment to sustainability—not through regulations imposed by the government, massive amounts of money contributed by international nongovernmental organizations (NGOs) or unrealistic approaches devised by academics with little or no experience in the Mexican Caribbean, but through a program owned, in every way, by the fishers and their communities—demands painstaking, ongoing, collective effort.

With the meeting of the Vigia Chico cooperative about to start, 30 men, most in cutoffs and tee shirts, and two women, the dark-haired younger one in a brightly patterned long skirt of finger-pleated cotton gauze, the other in jeans, a blue bandana securing her blond ponytail, begin taking their seats around picnic tables. One of the men wears a white guayabera over black cotton trousers; another sports the uniform of a park ranger, but with the top two buttons of his navy blue shirt undone. Latecomers stop at a cooler to grab a drink from the selection of Sol longnecks, bottles of orange Fanta, and bright red cans of Coke, depositing coins in the glass jar set on the folding table nearby. Everyone settles in around the picnic tables or drags up a molded plastic chair.[20]

As the shuffling and murmuring die down, the cooperative's president, a deeply tanned man with salt-and-pepper hair and a neatly trimmed moustache, claps his hands loudly, calling to order the assembly of the Vigia Chico Fishing Cooperative. His tone cordial, he introduces the guests: the blond woman representing an international NGO; the park ranger, seemingly familiar to everyone present; the state government official in his guayabera; and the dark-haired woman and her colleague, a fellow of about 25 with collar-length brown hair, both graduate students at a university in Florida. The students are here, the president explains, to study the migration patterns of the spiny lobsters that provide the income, either directly or indirectly, for everyone in this town of 400.

The president reminds the group of the rules: Speak one at a time, respectfully, no interrupting. Any suggestions for changes in policies of the cooperative or in regulations governing fishing must come from the fishermen themselves. The representatives of the government agencies and the Sian Ka'an Biosphere authorities are there to listen and to answer questions; otherwise they will speak only if a suggested change would violate an existing law or directive. The woman from the NGO is there to observe.

"Her organization wants to use us as a model for sustainable fishing programs in Africa and the Pacific, including our bottom-up style of management," the president announces. "We should all be proud."

Gesturing to the young man and woman from the university in Florida, he continues: "These young scientists would like you to allow them to place tags on some of the lobsters in your *casitas cubanas* now, during the off season, and have you report any tagged lobsters you harvest after the season opens in July," he says. "That way, we will know where our lobsters go when they are not at home."

"*Casitas,*" Spanish for "little houses," are artificial shelters the fishermen place on their *campos,* Spanish for "fields," the allotments granted to individual lobstermen (in this part of Mexico, commercial fishing remains a man's job) or their families. Lobstermen in Quintana Roo often call the shelters "*casitas cubanas*" because they were first employed in Cuba. In the Sian Ka'an Biosphere Reserve, the *campos* are laid out along the Bahía de la Ascensión and the Bahía Espiritu Santo.[21] Mexican federal law forbids outright ownership of underwater property, but exclusive concessions to harvest a given plot can be held and even passed on to heirs. In Sian Ka'an, the fishermen do own their *casitas,* and they can put as many on their allotted *campos* as they want.[22]

Ranging in size from slightly more than a square mile to slightly less than eight square miles,[23] the *campos* avoid the "tragedy of the commons"[24] articulated by British economist William Foster Lloyd in 1833: If a resource, such as pastureland, is held collectively, individuals will exploit it for their personal benefit, ignoring the good of the community.[25] The fishermen have incentives to steward their *campos,* avoiding overfishing them, keeping them free of trash, and investing in *casitas.*

In Cuba the *casitas* are known as *pesqueros* and often made of palm trunks, recycled hardwood, or scrap lumber. The *casitas* used in the Mexican Caribbean consist of a square or rectangular concrete slab measuring one to one and a half yards on a side, propped up three to six inches by concrete blocks, PVC pipe, or pieces of wood, and open along at least one edge.[26] *Casitas* differ from conventional lobster traps. Baited, traps have a one-way door and a rope so that fishers can simply haul them on board their boats. The wood-and-wire traps make harvesting lobsters easy, but they do nothing to recruit young lobsters or encourage their growth into commercially and reproductively valuable specimens. However, lobster fishers in the Florida Keys like this "traditional" equipment so much and have been so successful at making their case for it that the Florida Fish and Wildlife Conservation Commission declared *casitas* illegal in Florida waters,[27] a status still in effect in 2020.[28]

Casitas are designed to give lobsters the sense of security they would get from a niche in the reef, where competition for commodious crevices can be intense.[29] Because the fishermen place the *casitas* on the sandy flats or seagrass beds in or near the bay, these artificial shelters provide additional real estate to entice more lobsters to the area. In particular, the *casitas* draw juveniles, which are especially vulnerable to predators. For larger juveniles, cohabiting with adults offers protection. For smaller juveniles, what might be a "safe house" can be a literal dead end, because *casitas* can also harbor predators.[30]

A *casita* still may offer better odds for survival than a little lobster would get on its own. Between 80 and 96 percent of juvenile lobsters become dinner for octopuses, crabs, and other predators,[31] but once they reach full size, spiny lobsters seem able to hide or defend themselves, except from people, and can live 15 to 30 years.[32] Gathered together, antennae out, adult lobsters are good at warding off predators, but for juveniles this defense strategy is less effective. By cohabiting with adults, juvenile spiny lobsters may benefit from the protection of their roommates' longer antennae.[33]

Deployed in the shallows, *casitas* enable lobsters of all sizes to hunt more efficiently, venturing out to snag their own prey and scampering back home before attracting the attention of something that might want to eat *them*.[34]

Banco Chinchorro has no *campos*, but since 2010 the cooperative there has employed *casitas cubanas* to increase the population of lobsters and the ease of catching them.[35] Along the fringe reef surrounding the atoll, crevices in the coral provide ready refuge for lobsters, but the 900 collectively owned *casitas* offer them options in the seagrass beds and sandy areas of the lagoon in the middle of the atoll.[36]

From the fishers' point of view, lobsters are a lot easier to harvest from *casitas* than from niches in the reef, a process that requires a lot of free diving as deep as 60 feet and often to 45 feet.[37]

To harvest lobsters from the *casitas cubanas*, fishers travel to their *campos* in *pangas*, open skiffs 19 to 28 feet long, with upswept bows for buoyancy and blunt sterns for securing outboard motors. The fishers go out in groups of three—one to drive the boat, the other two to snorkel. The details of the bottom let them know when they've reached the *campo* assigned to one of them or to their family. As the two snorkelers jump into the water, one holds what looks like a butterfly net. Diving down to a *casita* set 12 or 15 feet below on the bottom, the first snorkeler pulls off the heavy concrete structure while the one with the net snags the lobsters inside. They bring them up to the boat captain, who measures them. Any that are too small or are egg-bearing females go back to the water.[38]

In January 2020 Omar Vidal, a marine biologist who headed the World Wildlife Fund Mexico for 14 years and was an independent consultant at the time this book went to press, spent several days in Punta Allen observing this method and pronounced it remarkably efficient.

"It was the first time I've tried the *casitas cubanas*," he told me. "It was amazing. We spent three hours, and they caught like 40 kilos of lobsters."[39]

First, the fishers take their catch to the cooperative's weigh station. Then they sell the lobsters to the buyers waiting on the dock, who in turn sell them to restaurants, seafood markets, and shippers.

A fisherman in Sian Ka'an can monitor the lobsters in his *campo* during the closed season, then, when the season opens, capture the full-grown adults with nets or loop snares. Unlike gaff hooks, snares allow lobsters to be harvested and sold alive, which increases their value.[40] When Kim Ley-Cooper studied Mexico's Caribbean fishing cooperatives for his 2015 dissertation, the price of lobster tails at the dock was US$38 per kilogram, while live lobsters brought US$20 per kilogram. Because whole live lobsters weigh three or four times the weight of their detached tails, using snares added more than 30 percent to the value of each lobster harvested.[41]

From the perspective of the environment, the relative ease of taking lobsters from *casitas* in the shallows discourages fishermen from capturing them on the Mesoamerican Barrier Reef, where even responsible fishing practices can damage the delicate coral ecosystem.[42]

Casitas aren't cages, and lobsters do move around. Large lobsters often live deeper—at depths below 60 feet, which in many jurisdictions are off-limits for fishing by regulation but would be impractical for fishers using snorkels anyway, since they are below the level anyone but trained free divers would go.[43] Surface water temperatures along the Mesoamerican Barrier Reef and the lagoons and bays adjoining it range from 76°F to 90°F; but in October through March, occasional north winds cause temporary drops in water temperature near the surface, triggering migration by juvenile and adolescent lobsters from the shallow bays to deeper water. Lobsters also relocate due to seasonal changes in salinity. During the summer rainy season, when mainland runoff dilutes the salt near shore, they make for the deeper waters at the mouths of the bays,[44] only to be lured back to the shallows by the greater abundance of prey.[45]

Lobsters can even travel internationally. Having four countries share the Mesoamerican Barrier Reef makes it more difficult to manage fisheries. Not only are there different national regulations and enforcement to deal with; but often fishers out on the water cross national boundaries;[46] and the lobsters have no way of knowing whether they're in Mexico or in Belize.

But back to the meeting: After wrapping up the introductions, the president of the Vigia Chico Fishing Cooperative pauses briefly. A voice from the back suggests mischievously: "Maybe these two scientists from Florida would be willing to tag the wives of some of my friends here."

A faint chuckle spreads across the room, accompanied by some jostling.

The president claps his hands again.

"This is a serious meeting," he says. "Our time is valuable. Even now in the off-season, each of us has a lot to do, scraping and painting our boats, mending our hand nets, overhauling our motors. Does anyone have a matter for the group to discuss, perhaps a resolution to propose for adoption?"

A man of about 30 rises.

"I think our cooperative should limit harvesting to snare loops and hand nets. No more gaffs and hooks. That way, Punta Allen can make its name as the capital of live, healthy lobsters."

When the president opens the issue for discussion, it becomes clear that a significant percentage of the fishermen object. They already operate under a lot of restrictions. At any rate, live lobsters always bring more at the dock, providing a clear incentive to favor snares and nets over gaffs voluntarily.

Everyone agrees to delay the suggestion for discussion once the season opens.

A middle-aged man halfway back rises, slapping his palm on the picnic table.

"The most urgent subject for us today is what to do about the middlemen," he declares. "The open season is two months away, and all of us are living on loans from these leeches. They charge us 20 percent against our future catches. This is legalized banditry! Once the season opens, they will own every lobster I catch during July and August. After that, I will be playing catch up, with nothing left over to save so that I could borrow less next year. But without the bloodsuckers' money, I cannot feed my family, maintain my *casitas cubanas*, keep my boat in shape, or put gasoline in the motor. We have to find a better way!"

Although this meeting scenario is an invention, the structure and function of the cooperatives, the rules they follow and the challenges their members face are real. Thanks to the *casitas* and the enlistment of the members of the cooperative in enforcing regulations against illegal fishing practices, the population of spiny lobsters in Sian Ka'an and Banco Chinchorro is gradually increasing, apparently through recruitment of juveniles to the *casitas*, where adults protect them from predators. But despite being organized into cooperatives, local fishermen struggle at the bottom of the economic pyramid when it comes to profiting from their labor.

Lobster fishers in Quintana Roo lack the resources to carry them through the closed season. Even once the season opens, hotels and restaurants often take as long as three months to pay for the lobsters they receive, so middlemen loan the cooperatives money secured by the next season's catch,[47] charging about 20 percent interest on these loans. Also, their sophisticated transportation and storage facilities, which the fishing cooperatives lack, enable them to supply the hotels and restaurants directly. By marking up the dock price 50 percent, the middlemen have plenty of cushion to withstand the slow payment policies of the hotels and restaurants, which, in turn, benefit by being able to charge diners US$100 per kilogram for lobsters sold for US$20 per kilogram at the dock.[48]

Two of the six cooperatives allowed to harvest lobsters from Sian Ka'an and Banco Chinchorro have been caught in a permanent cycle of debt, owing so much that their annual income barely covers their members' living expenses, the maintenance of their *casitas* and boats, and the money they spend on gas and equipment during the open season. The carrying costs have left the cooperatives and their members without savings to invest in the following season's operations, forcing them to borrow even more from middlemen who are willing to make high-interest loans with unfished lobsters as collateral. Normal Mexican banks charge yearly interest rates of 12 percent or less but demand that the borrowers put up buildings, boats, vehicles, and other hard assets to secure the note. Unable to afford to risk these and thus unable to get bank loans, the fishermen have to accede to the high rates charged by the middlemen, effectively dropping the value of next season's catch by 20 percent, a bite that renders the fishery unsustainable economically over the long term.[49]

But what if the spiny lobsters sustainably harvested by members of the six fishing cooperatives could be made more desirable to the market? What if diners in restaurants and shoppers in seafood markets could ask for them specifically and be willing to pay more?

The cooperatives voted to adopt "Chakay," Mayan for "lobster," as their label, like the *appelation controlée* for French wines. The idea wasn't new to Mexico, where every bottle of liquor labeled "Tequila" is made from agave harvested near the eponymous town in Jalisco.[50] "Langosta Chakay," followed by "from the Banco Chinchorro and Sian Ka'an Biosphere Reserves," indicates the provenance of the lobsters when they reach the market.[51] More than that, it testifies that the lobsters sold under this label were harvested sustainably, under conditions that were both ecologically sound and socioeconomically fair.

The idea behind eco-labeling is to add value to the catch, so that the fishers will be able to maintain or increase their incomes without having to fish harder or deplete the population of lobsters.[52] "Langosta Chakay" became the first eco-label registered with the Mexican Institute of Industrial Property; the label belongs to the "Integradora de Pescadores de Q. Roo," the six fishing cooperatives that hold exclusive rights to fish commercially in the two bioreserves. These cooperatives also became the first lobster fisheries in the world to be certified ecologically sustainable by the Marine Stewardship Council (MSC). Founded in 1999 by the World Wildlife Fund and Unilever, the Dutch-British food and cosmetics giant, the MSC is now an independent international nonprofit organization that has certified hundreds of fisheries around the globe.[53] (After several years of applying for and being granted MSC certification, the lobstering cooperatives ruled that the process was too expensive, opting to follow the standards set by the MSC with support from the Mexican NGO Comunidad y Biodiversidad [COBI], but without the official MSC certification.)[54]

The hope is that restaurants will proudly list the provenance of the lobsters they serve and that consumers will be willing to pay more for lobsters thus sourced. Part of the concept is that the "Chakay" eco-labeling will change the way the local fishers catch lobsters, including the gear they use, for example, replacing gaffs and hooks with loop snares and hand nets.[55] Even without the eco-label, live lobsters are worth more than tails.

The Sian Ka'an and Banco Chinchorro fishermen are pushing for more visibility for the "Chakay" eco-label and its promotion to ultimate consumers, especially for marketing strategies differentiating their sustainably caught lobsters from those harvested and sold illegally,[56] so that the cooperatives will become the logical source for restaurants and markets along the Riviera Maya and beyond.[57] Armed with the prestige and 33 percent value added by the "Chakay" eco-label, the cooperatives are seeking to secure sources of working capital at more reasonable rates—6 percent to 13 percent.[58] (The existing middlemen are fighting these efforts, because they currently hold the power to drive down the prices they pay at the dock.) In the process, all the stakeholders in the commercial supply chain must take moral responsibility for ensuring that a fair share of the value added by eco-labeling makes its way to the fishermen and their communities.[59]

Ultimately, it's up to us, the end consumers, to demand that all the seafood we consume is sustainable—for the species, for the ecosystem, and for the villages that depend on it for their livelihoods and for their ways of life.

18

Thinking Globally, Acting Locally— and Collaboratively

In March 1999, five young Mexican conservationists got together to form a nonprofit they called Comunidad y Biodiversidad (Community and Biodiversity; COBI). Concerned both about the health of the rocky reefs and kelp beds of their country's west coast and Gulf of California and about the growing poverty in developing countries, they also admired the diversity of the marine life in the oceans around Mexico and Central America. They figured that promoting healthy reefs and kelp beds could help relieve poverty by supporting sustainable small-scale fisheries and the communities that depended on them and by arranging economic incentives for conservation. From the start, COBI's approach has been to engage communities at every step, from identifying priorities to setting the boundaries of Marine Protected Areas. One arm of the organization's mission is the development of local leadership.[1] Since 2014, COBI's efforts have attracted more than a million dollars in grants from the David and Lucile Packard Foundation.[2]

COBI initially focused on the Gulf of California's Midriff Islands.[3] Expanding first in northwest Mexico, in 2008 the organization initiated a program in Quintana Roo. The local fisheries in the southern part of the state were already well-organized. Politically, commercial fishing on Mexico's Caribbean coast had been something of a stepchild. Although Mexico's Pacific coast had its own share of local fishing cooperatives, it was also home to industrial-scale fisheries big enough to garner attention from the national government in Mexico City. But while the Pacific fisheries were more productive, in terms of total catches, commercial fishing in Quintana Roo employed more people.

"They had to organize themselves, because no one else was going to do it," explained Stuart Fulton, COBI's coordinator of marine reserves, when I spoke with him by phone in January 2020.[4]

Joining a local fishing cooperative is hard. The rules vary from one to another, but all of them restrict membership. Some require a vote by existing members. For others, membership can only be inherited.[5]

Cooperatives work best in remote areas, where fishers rely on them, for example, to obtain gas for their boats. That's why the fishing cooperatives are stronger in southern Quintana Roo than in the northern part of the state, where highly developed tourism in Cancún, Cozumel, and Playa del Carmen reduces the incentive to work together. "When there's a hotel on the beach and you sell them your catch for cash, you may make the same amount of money, but the cooperative loses money," Fulton said.[6]

But the booming tourist industry in northern Quintana Roo presented its own opportunities for conservation. COBI figured that the region's larger businesses would recognize the benefit of protecting the reef. The "*comunidad*" part of its mission could involve everyone from fishers to restaurant waitstaff to taxi drivers to hotel managers.

"Almost all businesses in the area are dependent on a healthy reef," Fulton told me. "Big hotel chains in Cancún are beginning to get eco-certified and have corporate responsibility programs. Boutique hotels are starting to serve sustainable seafood."[7]

Another reason COBI opened an office in Puerto Morelos, about 18 miles south of Cancún, was the unique beauty and environmental value of the reef itself. "The Mesoamerican Barrier Reef is an iconic ecosystem," Fulton said.

When he first arrived at COBI's new office, Fulton set about working with the local fishers to select the boundaries for no-take zones. A certified dive instructor, he also trained the fishers to scuba dive and to collect data on the commercial species and on the health of the reef in general for annual surveys.

The fishers have proven good at their new job. "The scientists still do best," Fulton noted. "But the quality of the data we're getting from the fishers is a similar standard to what we get from the graduate and undergraduate students."[8]

COBI passes the results along to the Healthy Reefs for Healthy People Initiative (HRI), which integrates the information into its annual "report cards" and uses it in its public education programs.

In 2004 the World Wildlife Fund, the Meso-American Barrier Reef System Project, the World Bank, the Nature Conservancy, the Summit

Foundation, and Perigee Environmental Consulting had come together to form HRI, the vision of Melanie McField, a former Peace Corps volunteer in Belize. McField was convinced that the health of any ecosystem and the physical, social, and economic health of the people who inhabit it are intimately entwined.[9]

Roger Sant, the businessman who established the Summit Foundation, posed a tough question: How did HRI plan to measure the success of its efforts to improve the health of the reef?

"At the time, scientists didn't agree on what that meant," McField explained. "He asked me to put together a group of scientists to establish quantitative benchmarks and targets, like you would in business, and be able to track the reef with some index like the Dow Jones Industrial Average.

"The scientists didn't like it, but they understood the need. We had a great meeting in Miami in 2004 to set the targets and discuss indicators."[10]

In 2007 McField, independent consultant Maya Gomez, Matthew McPherson of the National Oceanic and Atmospheric Administration and Patricia Kramer of Perigee Consulting compiled the results into the *Guide to Indicators of Reef Health and Social Well-Being.*

As of 2021 HRI operated with a staff of six—McField, a communications consultant, and country coordinators in Mexico, Belize, Guatemala, and Honduras.[11] HRI also collaborated with 70 other environmental organizations as "a catalyst to improve our collective conservation impact in the Meso-American Reef Ecosystem." The Belize Audubon Society, which celebrated its fiftieth anniversary in 2019, participates in the HRI surveys[12] as does the Southern Environmental Association, which uses trained volunteers, known as "citizen scientists," to monitor Belize's Marine Protected Areas at Laughing Bird Caye National Park and at the combined Gladden Spit and Silk Cayes Marine Reserve, where whale sharks congregate to feed on the spawn of snappers.[13] "The good thing about Healthy Reefs is that we have this alliance, and we meet every year to discuss what is going on in the other countries," explained Marisol Rueda Flores, HRI's communications consultant.[14]

HRI conducts a biannual ecological audit of the reef in each of those countries, using the same criteria in each. These audit group 28 recommended actions into seven general themes: Marine Protected Areas; Ecosystem-Based Fisheries Management; Coastal Zone Management;

Sanitation and Sewage Treatment; Research, Education, and Awareness; Sustainability in the Private Sector; and Global Issues. This multinational assessment enables HRI to identify trends and results country by country, as well as for the Mesoamerican Barrier Reef as a whole. A given country may lead in one recommended action and trail in another.[15]

The eco-audit also serves as the basis for a spiffy user-friendly summary dubbed the *Healthy Reefs Initiative Report Card*. When held one way, the *Report Card* reads in English; turned over, it reads in Spanish. In Guatemala, where the illiteracy rate hovers around 20 percent,[16] HRI prints a chart with pictures of the species being monitored.[17]

The brochure also focuses on communicating the connection between reef health and human health to policy makers, decision makers, and the public.[18] Information transfer among stakeholders presents an ongoing challenge.

"You need the means to communicate the science to politicians, to take those papers that are stacked on shelves in universities and communicate their findings to the policymakers," Roberto Pott, former HRI in-country coordinator for Belize, told me.[19]

With support from the Palo Alto–based Gordon and Betty Moore Foundation, Conservation International, a global NGO, has produced a publication aimed to help scientists learn to do that—and to help politicians and other decision makers in tropical developing nations explain the realities of policy making to scientists. *A Decision-Maker's Guide to Using Science/A Scientist's Guide to Influencing Decision-Making* focuses on marine resources. Like HRI's bilingual Report Cards, it reads two ways: Held one direction, it addresses scientists. Held the other, it speaks to decision makers.

"Creating social change and solving environmental problems require both knowledge and power," the brochure declares. "Scientists have knowledge, but typically limited authority to change behavior. Decision-makers have power, but may lack in-depth knowledge of particular problems."[20] The dual guide explains that scientists typically focus on one area of expertise, while decision makers are generalists.[21] Science begins with a planning phase and moves on to data collection and analysis and finally dissemination of results.[22] Decision making, on the other hand, starts with issue identification and proceeds to implementation.[23] The guide urges scientists to synthesize existing scientific findings so that decision makers can understand and apply them.[24] It also recommends that 15 percent of the budget for any research project be earmarked for communication to decision makers and the public.[25]

Like Conservation International, the Healthy Reefs for Healthy People Initiative places a priority on communicating the connection between reef health and human health to policy makers, decision makers, and the public.[26] The 2010 *Healthy Reefs Initiative Report Card* recommended that 20 percent of a country's territorial seas be designated as Marine Protected Areas (MPAs), limiting fishing, diving, and other human activities. By 2018, 57 percent of these waters lay in MPAs. Although fishing was totally prohibited in just 3 percent, these included spawning areas for such species as Nassau groupers. In Belize the percentage of its waters in no-take zones was 4.5 percent. In 2019 the country expanded that to 11.6 percent, including deep sea waters ranging from 600 to 9,000 feet.[27]

HRI monitors found four times as many adult fishes in MPAs where fishing was banned as in areas that were partially protected but where fishing was allowed.[28] Even more impressive, the biomass of snappers and groupers was 10 times greater. The number of large groupers was especially impressive in long-established Marine Protected Areas (MPAs).[29]

Such MPAs work, increasing biomass an average of 446 percent. One portion of the Mesoamerican Barrier Reef half a mile from Laughing Bird Caye National Park in Belize went from "critical" to "good" within just three years.[30] In Guatemala, the first three no-take zones were an initiative from the community, but there was nothing in Guatemalan laws to establish the protected areas. The fisheries and environmental groups had to work with the government to get the regulations on the books.[31]

Some local programs are garnering international attention. One, Puerto Morelos–based Guardianes del Arrecife (Guardians of the Reef), has the support of the Nature Conservancy. Guardians began modestly. Local fishermen noticed a decline in the health of the corals in the Puerto Morelos Reef National Park offshore. Because it was now a no-take zone and thus off limits to commercial fishing, the fishermen had become tour guides, making their livings by showing the reef to the visitors who flocked to Cancún and Playa del Carmen. The fishermen-turned-guides appealed to the National Commission of Aquaculture and Fisheries (CONASPECA) for scientific and technical assistance in halting the deterioration.

"They now relied for their incomes on taking tourists to the beautiful reef," explained marine biologist Gustavo Guerrero Límon, who was recruited to train locals to do coral restoration. "If they lose the corals, they lose their jobs."[32]

Essential to the program was "passive restoration"—training the local guides to avoid touching anything underwater and to insist that the divers and snorkelers they took to the reef do the same. With the cooperation of the community, some sections of the Mesoamerican Barrier Reef have been declared "core zones," where even diving is prohibited.[33]

With the help of experts from the Nature Conservancy, volunteers with Guardians of the Reef have also learned to repair corals damaged in storms using a special cement to reattach the broken bits.[34]

Another strategy, this one in Belize, enlists the country's engineers to help protect the reef and the communities that depend on it. With support from the Inter-American Development Bank, in 2019 the University of Belize Coastal Resilience Program conducted a three-month seminar in designing coastal infrastructure to withstand the impact of climate change.[35]

Belize clearly needs its engineers' enhanced expertise. Of 167 developing countries, Belize ranks eighth in climate change risk as assessed by the Green Climate Fund, established in 2010 by the United Nations Framework Convention on Climate Change to provide economic support for projects to combat the negative effects of climate change. The assessment included the other three countries on the Mesoamerican Barrier Reef—Guatemala, Honduras, and Mexico—but GCF deemed Belize to be most at risk. Between 1990 and 2008, climate change cost the country an average of almost 4 percent of its gross national product per year.[36]

Belize's NGOs have stepped forward to address the challenges. As mentioned previously, for more than 50 years, the Belize Audubon Society has been a leader in environmental activism in the region and a model worldwide. More recently, the Southern Environmental Association (SEA Belize), focusing on the districts of Stann Creek and Toledo in the country's south, has demonstrated that local fishers, dive and nature guides, and other "citizen scientists" can be effective at monitoring the health of the reef and educating their communities in methods for protecting it.[37]

"Adaptation should be one of our first priorities," Omar Vidal explained, "adaptation in terms of the reef and adaptation in terms of the people who live here. If we protect the reef, the estuaries, and the wetlands, we will be protecting the lives of the people."[38]

Vidal acknowledged that climate change was the most insidious threat to the Mesoamerican Barrier Reef and marine ecosystems worldwide.

"It's something we will not be able to stop in the long term," he admitted. "We need to help nature to adapt. One of the ways we can do this is to stop threatening the reef with various human activities—pollution from agricultural runoff, sewage, and tourism."[39]

Among the warning signs of destructive human impact, new bacteria are attacking stressed corals, and huge mats of sargassum seaweed are appearing, far more than the periodic influx of previous decades. Not only does the sargassum damage the reef, it also hurts the fisheries. And that exacerbates another insidious threat: poverty.[40]

Through programs to reduce runoff, the World Wildlife Fund (WWF) is helping address both the health of the reef and the economic health of the people along it. WWF programs teach sugar farmers to replace toxic pesticides with biological methods of pest control and chemical fertilizer with compost. The Fund also works with local communities and Indigenous groups to keep critical landscapes intact, retaining fresh water on their farms where it does good, rather than allowing it to wash down to the reef, where it does harm. On the coast itself, WWF is training schoolchildren to replant mangroves in the areas where they have been torn out.[41]

On December 5, 2016, the Mexican federal government declared 14 million acres along the coast of Quintana Roo the Mexican Caribbean Biosphere Reserve, making it the country's largest protected area. This matchless expanse of reef, lagoon, and wetlands included the Sian Ka'an and Banco Chinchorro reserves. The declaration notably excluded the burgeoning tourist destinations of Cancún and Playa del Carmen, but it did extend protection to Cozumel and nearby Isla Mujeres. Although commercial fishing and resort development were banned in only certain places, the biosphere reserve designation gave the authorities the tools to manage human activity along most of the state's coast for the benefit of the inhabitants and of the Mesoamerican Barrier Reef. The German government contributed 10 million euros to the effort[42] and is providing support to NGOs working with local groups in Quintana Roo and in Guatemala to help assure that the social and economic benefits reach the communities that rely on the reef.[43]

Omar Vidal argued that the large national and multinational corporations that profit from the reef's natural beauty should be the primary sources of funding for preserving it. "The cruise companies and the hotels and resorts along the reef are the ones that benefit financially from it," he explained, "so they should be the ones paying the local NGOs and cooperatives to protect it, and they should be helping provide a decent standard of living for the people. Tourists come and go, but the community stays there."[44]

How well the Mexican Caribbean Biosphere Reserve will operate in practice, given the economic and social pressures endemic to the region, is a question for the coming decades to answer. "Up to now, the decree is just on paper," HRI's Marisol Rueda Flores told *The Guardian* in December 2017, noting that at the same time the government was establishing the reserve, it had cut funding from the agency overseeing marine parks. "Mexico has some of the best laws for protecting coastal areas—the problem is no one follows them. The government has barely begun to realise [*sic*] the urgency of this situation."[45]

Meanwhile, at the 100 sites monitored by the Healthy Reefs for Healthy People Initiative, the value of coral cover increased 38 percent between 2006 and 2016. Granted, climate change continues to take its toll. Rising ocean temperatures and acidification, fiercer storms, more runoff from the land—all these impact this unique ecosystem, as they do marine environments worldwide. However, HRI's 2018 Report Card concluded that collaborative efforts along the Mesoamerican Barrier Reef seem to be helping corals, and the entire ecosystem, become more resilient in the face of these global threats.[46] "We may not be able to do much about the damage caused by climate change," Melanie McField conceded, "but we can minimize the other factors that impact the reef."[47]

Boston University marine biologist Les Kaufman and seagrass scientist John Tschirky agreed. "Coral reef communities can be more resilient to acute local and global scale impacts if they have a healthy complement of microbes, corals and other invertebrates, and herbivorous and predatory fishes," they noted in a 2010 monograph published by the Science and Knowledge Division of Conservation International.[48]

Because macroalgae continues to smother sections of the reef, by 2018 HRI was looking at ways to reintroduce spiny sea urchins and Caribbean king crabs, both effective grazers. By that time, Belize and Guatemala had laws on the books protecting another enthusiastic algae-eater, parrotfishes, as did Honduras for the Bay Islands that form the southern tip of the Mesoamerican Barrier Reef. Despite the complication that parrotfishes were a traditional food source for Indigenous groups on its West Coast,[49] Mexico instituted a similar ban in 2019.[50]

On June 5, 1997, World Environment Day, the leaders of the four countries along the Mesoamerican Barrier Reef had signed the Tulum Declaration pledging support for protecting it,[51] but this official cooperation had proved challenging in practice. One reason was that these countries themselves were, and are, culturally diverse. A traditional food fish among the people in one district may be off the menu in another. One group may use a particular marine species in its ceremonies, which another group considers alien.

Another challenge has been the difference in population size and political influence. Belize is the only one of the four countries with its coast entirely within the Mesoamerican Barrier Reef ecosystem, but its estimated 2021 population was just 402,154.[52] Immediately to the north, the east coast of the Mexican state of Quintana Roo also lies entirely along the reef. Although Quintana Roo's 2020 population of 1.7 million was more than four times that of Belize, it was only 1.4 percent of the population of Mexico, which topped 130 million, most concentrated in huge cities in the interior.[53] Although Guatemala had 18.25 million inhabitants in 2020, only 430,000 lived on or near its relatively short stretch of Caribbean coast.[54] In Honduras, with a total 2020 population of more than 10 million,[55] fewer than 70,000 inhabited the Bay Islands of Roatán, Guanaja, Útila, and the neighboring smaller islands and cays according to the most recent figures available (2015).[56]

Tensions between these countries can also complicate collaborative efforts to steward the international treasure they share. Melanie McField explained in an e-mail communication in February 2021: "One of the challenges that Belize has always felt is that the other countries are larger and more powerful and more prominent on the world stage. And then there's the issue of Guatemala claiming over half of Belize. This dispute has become more of an issue lately and is headed to the International Court of Justice. Over the years, the dispute has gone up and down, complicating the regional approach to conservation."[57]

Since well before Belize gained its independence from Britain, Guatemala has claimed all or part of the former colony. A road map of Mexico that I bought at the Mexico City airport in the 1990s featured a map on the reverse side showing Guatemala extending east to the Caribbean all the way from Mexico to Honduras, with no international border at Belize.

In 2016 Guatemalan fishermen out looking for sharks in the deep water reported to HRI's Guatemalan division that they had "found some rocks." The "rocks" turned out to be a previously unrecorded reef, about nine miles long and two miles wide, located seven miles from the coast and rising within 15 feet of the surface. Given the name Cayman Crest, because it was near the Cayman Trench, the reef was a pristine, beautiful coral spur-and-groove structure, with walls dropping to 900 feet, gin-clear water and large snappers and groupers.[58]

But the new reef lay between Guatemala and Belize, south of the Sapodilla Cayes. "That's the difficult part about the site," Ana Giró Petersen, HRI's in-country director for Guatemala, told me in a 2016 phone interview. "The problem is now there's political tension between the two countries, so we have to work strategically. A binational park or a binational protected area would be an enormous accomplishment."[59]

Doing its best to by-step the boundary dispute, the Healthy Reefs for Healthy People Initiative set about mapping Cayman Crest and studying the ecosystem. That brought up another challenge: the scientists studying the new reef had to drive six hours from Guatemala City to Puerto Barrios, then undertake an hour-long boat ride. HRI decided to start integrating people from the nearby communities into the research. A partner organization took on training and certifying locals in scuba diving. In addition to helping monitor the site, increased community involvement would make it *their* reef. After all, the fishermen who first reported the "rocks" already felt a sense of connection and trusted the organization to make good use of the information, which was a promising sign that HRI's community partnerships were working.[60]

Ian Drysdale, HRI's in-country coordinator for Honduras, told me: "One of the key success factors for the Healthy Reefs Initiative has been working with everybody—the local community, the NGOs. Anybody who wants to can become part of the team, and that's created a great sense of family. We get all the partners together in the four countries along the Mesoamerican Barrier Reef. Size doesn't matter. What matters is our perseverance and inclusion. We have this big table where they can all sit down. They can inspire each other and also point fingers at each other and say, 'Hey, you're not doing your job,' or, 'Here's how we solved that problem you're dealing with.'" He added: "Having them be part of gathering and analyzing the data gives them a tool that they can use with their governments."[61]

In the Bay Islands of Honduras, one of the biggest challenges to the well-being of the Mesoamerican Barrier Reef was the lack of federal funding for management of protected areas. HRI and its community partners responded with something it called the Patrol Initiative. They acquired a boat, then acquired a motor, then hired a captain to drive around the coast of Roatán to spot who was doing illegal fishing. In Honduras, Drysdale explained, what matters is *how* you fish, not what species you take.

"If you catch a huge grouper on a hook and line, you can keep it," he said. But nets, traps, and spear guns (except the distinctive yellow Hawaiian slings used for lionfish) are forbidden. Because nets and traps don't discriminate, banning them benefits all species.

And restrictions also benefit recreational fishers and guides. "We see more and more people come to fish in this part of the island, because that's where the big fish are," Drysdale told me.[62]

The Patrol Initiative captain took local police with him and the police made the arrests and confiscated the nets, traps, and spear guns. Eventually there were six boat captains patrolling the Roatán Marine Park, part of the Bay Islands Marine National Park, which extends 12 miles out from the coasts of Roatán, Guanaja, and Útila,[63] giving it a total surface area of 182 square miles.[64]

"Because Roatán is a small community, our boat captains are related to our fishermen or are married to their cousin," Drysdale noted, "so having the boat captains conduct the enforcement prevents resentment."

Federal and district governments, local municipalities in the area and NGOs interested in protecting the reef collaborate to manage the Roatán Marine Park. Much of the funding for this grassroots effort comes from the local dive shops, whose owners thought, "If I don't protect this, I won't have a healthy reef to promote." Everyone who dives around Roatán pays a fee to the marine park. Officially, the fee is voluntary, but most operators won't take a diver out unless he or she pays it. Each month, the park publishes its books, so anyone can visit headquarters and see how the money has been used.[65]

Like Honduras, Belize finds cost a challenge in enforcing regulations in Marine Protected Areas. "Marine enforcement is very expensive," Roberto Pott, former Belize in-country coordinator for HRI, said. "It takes a lot more gas to patrol a marine preserve in a boat than it does to patrol a preserve on land." Also, he added, patrols have to "catch them [people fishing illegally] in the act. That's difficult, especially at night."[66]

With the patrolling program demonstrating its effectiveness, the Honduras division of the Healthy Reefs for Healthy People Initiative began tackling a problem that Drysdale admits was "not as sexy or attractive"— sewage and sanitation. Untreated sewage quickly makes its way to the reef, where the nutrients feed the macroalgae that smothers the corals. At the same time, untreated sewage spreads waterborne diseases. An initiative that kept human waste from reaching the sea would provide a clear example of the interconnectedness between healthy reefs and healthy people. So Roatán built a new wastewater treatment plant and set about connecting every home and business on the island. Initiatives like this that have clearly defined, realistic, measurable objectives appeal to NGOs and to philanthropies. For example, the Coral Reef Alliance (CORAL), a global nonprofit uniting scientists with fishermen, divers, and communities, partnered with HRI on this one, and in 2019 the Washington-based Summit Foundation gave it a one-year grant of $120,744.[67]

It helped that Drysdale served on the local water board, responsible for managing the new facility, as well as for providing potable water to everyone. Between 2013 and 2018, about 300 homes and businesses hooked up to the new wastewater treatment plant on Roatán's West End, preventing 30 million gallons of raw sewage from being dumped on the southern section of the Mesoamerican Barrier Reef each year and reducing the level of fecal bacteria by 95 percent.[68]

The collaborative effort also strengthened community partnerships. "We've found that by empowering local water districts, they can become stewards of the reef in managing the water," Drysdale said. "Even in our community, there was no consciousness about the importance of water. When I took this position on the board, there were no water meters in place, so there were lots of leaks and waste. One of the first things they did was install water meters and start charging people by the gallon, rather than a flat fee, for water. We reduced water usage by 25 percent."

The water board also offered the owners of residential and commercial property voluntary plumbing inspections. Although the board didn't pay to fix the leaks, the inspections were free, and the owners now had an economic incentive to make the repairs.[69]

In 2018 the Healthy Reefs Initiative Report Card declared that the Bay Islands now had effective wastewater treatment regulations.[70] Marine biologist and Emmy Award–winning documentary producer Sylvia Earle, founder of Mission Blue, joined HRI on a diving research expedition to

Honduras and got to see some of these reefs up close. During that trip she expressed optimism with regard to the conservation efforts along the Mesoamerican Barrier Reef: "You must be doing something right, because here, there are plenty of reasons for hope," she declared. "Cordelia Banks, off Roatán, Honduras, is one of the best places I have seen, even counting 50 years ago, an amazing stand with acres of staghorn coral."

Earle explained that she really likes the term "Hope Spot" as a concept that suggests that "places like these that are protected are really a cause for hope."[71]

Located next to Roatán's airport and between the towns of French Harbor and Coxen Hole, the Cordelia "Hope Spot" has 70 percent coral cover. "And it's the endangered species, staghorn," said Jennifer Myton, assistant program director for CORAL. "Researchers are studying why the reef is doing so well. It's swept clean by current, and it's recently been declared a protected area. The staghorn has created a little nursery, because there are lots of nooks and crannies where the small fish can hide. Monitors have been able to see an improvement in fish biomass."[72]

Setting aside part of a Marine Protected Area as a no-fishing zone allows juvenile fishes to grow in a protected environment. When they reach adulthood, some migrate to areas where fishing is allowed, a phenomenon termed "spillover." Thus, prohibiting fishing in one area can result in a larger catch for commercial fishermen in another area nearby. Half Moon Caye on Belize's Lighthouse Reef is an example of a preserve functioning in this manner.[73]

When it came to establishing no-take zones near the Cordelia Reef, CORAL, HRI, and other NGOs worked with the local fishermen to establish the boundaries. Once the fishermen understood that if they avoided fishing in some areas, their catches would improve in others, they agreed to collaborate, and the result exceeded everyone's expectations.[74]

Myton told me: "The average of live coral cover on the Mesoamerican Barrier Reef is 19 percent, so that when you see areas of 70 percent, that's remarkable. In this area, there is also the highest concentration of long-spined sea urchins. It's not really known why these survived, whether they were stronger or it was just because of a fluke of currents. Long-spined sea urchins are so important, because as a key herbivore, they keep the micro-algae at bay."[75]

But despite progress in Honduras and Mexico, sanitation, sewage, and runoff remain key issues for the Mesoamerican Barrier Reef. "I grew up in Belize City, and we used to be able to jump off the pier," Roberto Pott recalled. "But kids can't do that now, because the water's too polluted. That's a huge problem when it comes to the health of the reef and to human health. There is some truth to the argument that some of the nutrients and some of the trash in Belize are coming from Guatemala, but we have to look at what *we're* doing. Belize is trying to improve their sewage and wastewater, which is way behind. It's not that we can tell the government, 'Stop development while we fix the problem.'"[76]

Development of hotels, resorts, and cruise ship ports often outpaces development of infrastructure for wastewater treatment, while dredging for ports and runoff from roads damages the reef directly, and tearing out mangroves destroys the natural filtration system for agricultural and rainwater runoff. All the surface rivers in this part of Central America drain into the Caribbean; on the limestone platform of the Yucatán, underground rivers carry untreated sewage, petroleum effluent, and fertilizer and pesticides to the sea.

Because fresh water is less dense than salt water, runoff floats on the surface. Although the fresh water doesn't touch the reef itself, the silt and pollutants it carries block the sunlight that the corals' symbiotic algae require for photosynthesis. As a graduate student at Texas A&M, chemical oceanographer Andrea Kealoha investigated the mysterious 2016 die-off of corals, sponges, sea urchins, and other creatures inhabiting the reefs at the Flower Garden Banks National Marine Sanctuary. Heavy rains had swollen rivers in Texas and Louisiana, spilling muddy fresh water into the Gulf of Mexico, and a mysterious mat of whitish gunk had spread across part of East Flower Garden Bank, 70 miles offshore at the border of the two states.

Kealoha concluded that by blocking sunlight, the runoff had set up a chain that literally smothered the reef. Without the influx of oxygen from photosynthesis, the animals on the reef consumed the remaining oxygen faster, and there just wasn't enough to go around.[77] On the northern part of the Mesoamerican Barrier Reef, underground rivers carry pollutants; but beginning at Chetumal, where the Rio Hondo marks the border between Mexico and Belize, the threat multiplies as surface rivers drain the mainland, conveying silt and toxins across the seagrass beds to the reef, making this delicate environment vulnerable to the same kind of damage Kealoha observed 800 miles north.

"Because hard corals form the foundation of the ecosystem, the heavy influx of runoff can be devastating for the whole ecosystem," Kealoha told me. Fish can swim away to find healthier waters, but they don't return if the oxygen level remains low.

"When the coral reef dies, there's no food, no housing to support the animals that live there," she explained. "The mobile animals won't come back if the corals don't recover."[78]

Furthermore, as mentioned previously, runoff introduces nutrient-heavy sewage and agricultural waste into the environment. Reef-building corals require a low-nutrient environment to thrive, and industrial and agricultural toxins poison all creatures sharing the reef, including humans.[79]

Aquaculture compounds the problem, destroying mangroves and seagrass beds to build fish and shrimp farms, then polluting the water with the animals' waste. The Mexican government actively promotes aquaculture, noting that the industry valued at US$776.1 million in 2014 "afford[ed] significant investment opportunities," especially in raising seafood for export. Although Mexican aquaculture is most developed in the western states of Sinaloa and Sonora and in Veracruz, on the Gulf of Mexico, Pro México, the agency charged with promoting Mexico's exports, also extols the potential of Quintana Roo, which including bays and inlets has 480 miles of coast, all of it no more than a lagoon away from the Mesoamerican Barrier Reef.[80]

"The Riviera Maya has been devastated by human activity," declared Healthy Reefs for Healthy People Initiative communications consultant Marisol Rueda Flores, who is based in Playa del Carmen. "Right now, we are growing so fast that the water treatment plants that we have are not enough, and not all the people are connected to the sewers. The federal norms for wastewater treatment work well for places like Mexico City in the middle of the country, but not here because of the geologic conditions."[81] Partnering with other NGOs, HRI began pushing for norms appropriate for the Yucatán Peninsula.[82]

Clearly, the survival of the Mesoamerican Barrier Reef ecosystem depends on more than protecting the 60 percent of the 180,000 square miles that lie underwater. Conservation International (CI) proposed establishing Marine Managed Areas, "a combination of land and ocean, where all human activities are managed toward common goals."[83]

To make this proposal specific, CI came up with a short list of recommendations for sustainable development along the Mesoamerican Barrier Reef:

Locating roads well back from the coastline and building bridges that allow passage for migrating fish and other marine creatures.

Using mangroves and other plants to reduce erosion and sediment.

Utilizing agricultural practices that minimize sediment and chemical pollution.

Encouraging ecotourism and homestays.

Harvesting fish and other commercial marine species in ways that maintain their populations and ensure the health of the larger ecosystem.[84]

John Tschirky and Les Kaufman studied the results of establishing Marine Managed Areas (MMAs) at various locations around the world. Broader in focus than Marine Protected Areas, MMAs are a form of ocean zoning in which some areas are off-limits to any extraction (e.g., fishing, mining) and others are zoned for certain activities but with restrictions (e.g., sports fishing, scuba diving).[85] MMAs have very long time frames, because it may take generations for the benefits to be felt, in terms of "increased diversity and abundance of native organisms and ecosystem resilience."[86]

Establishing an MMA requires collaboration among local communities, business interests, government agencies, NGOs, and academic researchers. The process of building these partnerships can result in as many benefits, and challenges, as the MMA itself.[87]

The Healthy Reefs for Healthy People Initiative's 2016 eco-audit identified management of the coastal zone as one of the most critical challenges in the region. Balancing economic development, secure livelihoods for locals and long-term ecological sustainability was a herculean task. Since HRI's first audit in 2010, all four countries sharing the Mesoamerican Barrier Reef had shown improvement, but too often the management plans were not followed in practice. Corals destroyed through dredging, mangroves torn out to make way for beaches of imported sand—such physical damage is permanent.[88]

Even Belize, which consistently earns HRI's top marks, allowed Norwegian Cruise Line to develop a cruise ship port on the company's private island, Harvest Caye, off Placencia on the country's southern coast.

"They took a cay that had sustainable tourism and made it a port, even though that was contrary to the established sustainable tourism plan," Roberto Pott complained. "There's now huge development and a huge waste management problem."[89]

Twenty of the world's 30 largest cities are located on or near oceans, and more than half the people on the globe live within 100 miles of a coastline.[90] Like HRI, Blue Ventures, CORAL, and other NGOs, Conservation International has adopted an inclusive approach to protecting the environment: From its initiation through planning and on to implementation, any program must include the people directly affected and success must be measured in social, cultural, and economic, as well as ecological, terms. Otherwise, the program will never work long term.

The days of North Americans and Europeans with their PhDs and their foundation funding showing up to tell locals what to do are over. No matter how well-meaning, colonialism is colonialism, and the communities that have lived for centuries with and from the Mesoamerican Barrier Reef know colonialism far too well to cooperate with measures imposed from outside. This reef, after all, is their heritage, as well as the world's.

Belize has been committed to conservation since colonial days. As mentioned in chapter 2, the government declared Half Moon Caye a preserve in 1928. Although that enactment was designed to secure the nesting area of the red-footed booby, it raised awareness of the importance of protecting the habitat of rare species. In 1996, Lighthouse Reef, the atoll that incorporates both Half Moon Caye and the Blue Hole, became a national park.[91] (Laughing Bird Caye, off Placencia, had achieved that status in 1991.[92]) Local fishermen set up a trust to establish the Hol Chan Marine Reserve near the bustling tourist island of Ambergris Cay. A board of directors manages the resources, so the reserve is not completely run by the government.[93]

"Belizean NGOs co-manage protected areas in Belize," Omar Vidal explained. "They enter into agreements with the government in which they monitor and patrol both the water and the land. These NGOs are actually an interface between governments and communities. They're teaching us that civil societies can work together with government."[94]

Fishers in Belize express a sense of pride and ownership of the Mesoamerican Barrier Reef. Melanie McField explained that because they founded cooperatives before the country's commercial fishing industry took off, they didn't have to go up against established major fishing interests, as the Mexican lobstermen described in the previous chapter have had to do.

"The commercial fishermen own their cooperatives," McField said. "They eliminated the middlemen—the fish buyers. The fishermen get their share a couple of times a year, with the last payment dependent on the profitability of the co-op, and there's a savings plan that helps them send their kids to college, for example. Fishermen are not always the best financial managers and overseers, but the system does work and gives them ownership."[95]

One reason Belize's fishers have been willing to take a sustainable approach to fishing is the absence of one of the temptations to overfish: local demand. Belizeans don't have to rely on the sea for protein. The Mennonite farmers in the north and west of the country provide chicken, eggs, and cheese and also grow a lot of rice and beans.[96]

Whether it's sustainable fishing practices, environmentally conscious placement of roads and bridges or effective sewage treatment, minimizing human impact helps reefs recover from events such as bleaching resulting from El Niños and the warming of the oceans.[97] Portions of the Mesoamerican Barrier Reef suffered bleaching in 1995, 1998, 2003, 2005, 2008, 2009, 2010,[98] 2016[99], and 2017.[100] Hitting globally, the 1998 bleaching, the worst of these events, resulted in the death of 15 percent of the world's corals.[101] Two or three years later corals in healthy reefs had recovered, but reefs that were struggling with local challenges, such as pollution from development and agriculture, still hadn't rebounded eight years later.[102]

Although monitors for the Healthy Reef for Healthy People Initiative feared that the 2016 and 2017 events might rival 1998, the most devastating to the Mesoamerican Barrier Reef to date, the majority of the damage occurred at the southern end of the reef, where as much as 80 percent of the coral around Roatán and Útila showed bleaching. A few sites off Guatemala were affected, and damage off Belize ranged from 33 to 59 percent. Only a few spots of unnaturally pale coral appeared off Quintana Roo.[103]

Bleaching draws media coverage, partly because the white patches show up so dramatically in photos and videos. This draws attention to the risk of losing the world's corals. A third of reef-building corals face extinction. The populations of elkhorn and staghorn corals, the two main shallow-reef builders, have dropped 80 percent since the 1980s. Boulder star coral, another reef builder, has decreased 50 percent. As the oceans warm and become more acidic, corals become more susceptible to disease. Also, warmer oceans generate stronger, and probably more, storms, in which violent wave action damages corals.[104] Fortunately, staghorn and elkhorn are the fastest-growing of the reef-building corals. They readily reproduce asexually through fragmentation.[105] When a storm surge knocks off a section, a new colony can establish itself if the fragment ends up in a suitable place.

But publicizing bleaching events has downsides. Many otherwise well-informed readers and viewers and even reporters confuse bleaching with death—not surprisingly, given that those once-colorful coral colonies look downright skeletal. Figuring that it's too late to do anything, especially if they believe that we're past the tipping point for global warming, people surrender to hopelessness.

In reality, localized initiatives such as wastewater treatment and sustainable fishing work. There are more large fishes, and they're bigger.[106] Despite hurricanes and major bleaching events, HRI has reported a gradual increase in the health of the Mesoamerican Barrier Reef. Belize leads the four countries in implementing HRI's reef management recommendations, but Mexico and Honduras are close behind, and Guatemala has scored over half.[107]

One of the persisting threats to the Mesoamerican Barrier Reef is the growth of macroalgae smothering the coral. The problem is two-pronged: the aquatic environment has an overabundance of nutrients for the macroalgae to consume but doesn't have enough grazers that like to consume the booming macroalgae. The macroalgae crisis serves as a prime example of what happens when an ecosystem gets out of balance.

Just as we humans collectively caused the imbalance, working together we can help correct it. Sewage treatment, strict development regulations (and their enforcement), and enlightened agricultural practices can reduce the nutrients that feed the macroalgae. Protection of parrotfishes and other grazers can help bring that burgeoning macroalgae back under control.[108] As mentioned previously, parrotfishes face a particular potential risk. Although they may not be part of the traditional diet among the people living along the Mesoamerican Barrier Reef, they could become more appealing to fisherman as a food catch if commercial species like snappers and groupers decline.[109]

"Now, more than ever, we need to increase our conservation efforts and reduce human-induced impacts," Melanie McField told me. "People outside the region don't recognize all the effort that is going into this. If we have a hope of saving the ecosystem, this is it."[110]

Building sewage treatment plants, relocating roads, patrolling Marine Protected Areas, even reintroducing long-spined sea urchins—all that requires money, sometimes more than local businesses, like Roatán's dive operators, can contribute. Stephanie Wear, coral reef expert for the Nature Conservancy, was looking at ways to involve major corporations. After all, food and cosmetics multinational Unilever had partnered with the World Wildlife Fund to establish the Marine Stewardship Council.

Take pollution. "An example would be a sewage treatment company," Wear told the Conservancy's magazine. "Maybe we partner with a company like that to make it more affordable to bring sewage treatment to countries that need it—protecting reefs and helping people."[111]

Roberto Pott, former Healthy Reefs Initiative in-country coordinator for Belize, explained that through a collaborative effort, Belize City had improved its wastewater infrastructure. "There are people who have seen that there is this need and that there is an opportunity to bring in private industry to work with us," he said.[112]

Whatever the form of outside support, those providing it should take care to avoid disrupting the bottom-up structure of community collaboration. HRI, CORAL, and other twenty-first-century NGOs are accustomed to acting as catalysts, not as saviors long on money but short on local knowledge. That's why Blue Ventures, a worldwide organization based in London, is studying collaborative efforts to reduce lionfish populations in Belize to see if these projects can serve as models for community-based programs in Madagascar.[113]

"We're evaluating different models for managing lionfish populations below the target threshold," Jennifer Chapman, Belize country coordinator for Blue Ventures explained. "Of those that are effective, we try to evaluate which have the most social benefit."[114]

Human beings are as integral to the Mesoamerican Barrier Reef as parrotfishes and sea turtles, and for the reef to thrive, the people who depend on the ecosystem must thrive also. With the encouragement of HRI, some residents of fishing villages along the Guatemala coast have developed *tiendas,* small shops catering to visitors drawn by the marine reserve, and even an eco-hotel in Estero Lagarto at the mouth of the Rio Dulce.[115] Local residents are seeing the financial benefits of low-impact tourism.

And the other stakeholders are learning that bearing in mind the impact on communities and letting their members assume leadership are essential to long-term success of programs to protect the reef. Only then will collaborations be able to address the social and economic factors that lead to overfishing and poor land use.

Collaborations among fishing cooperatives, coastal residents, government agencies, scientists, and NGOs demand time, communication, and cultural sensitivity. In some times and places, these cooperative efforts work better than they do in others. But no one seems to find them controversial. The response is different when it comes to coral restoration.

At the time she died of complications of surgery for diverticulitis on October 25, 2018, at the age of 56,[116] University of Hawaii marine biologist Ruth Gates ranked among the most prominent of the world's scientists selectively breeding corals to resist rising ocean temperatures and acidification. She was president of the International Society for Marine Studies.[117] The question she wrestled with was why some coral reefs that had been given up for dead bounced back, while others didn't.[118]

Gates told *New Yorker* environmental writer Elizabeth Kolbert that she had concluded that the answer to saving coral reefs wasn't going back to a period with cooler, more alkaline waters, but moving forward into a future where human investigation would help counteract human-caused damage. Gates's research team would grow some corals in water as warm as the oceans are predicted to become, some in acidified water and some in water that was both warm and acidified. By breeding the best-performing corals to each other, Gates hoped to produce "super corals" that would do even better.[119]

The approach was not unlike agricultural corporations genetically engineering strains of corn that could withstand high doses of pesticides. Although no one was planning to make the oceans warmer and more acidic, humans just weren't doing enough to stop the process.

And that uncomfortable truth is at the root of the controversy. By selectively breeding "super corals" able to withstand rising temperatures and acidity, are scientists like Ruth Gates spreading, however unintentionally, the message that human beings can keep doing whatever we want to the planet, and our brilliant researchers and high-tech wizards will figure out ways to clean up the mess?

It may be the case that we need to act on all fronts, and quickly. If we don't bring ocean temperatures and acidification under control, we may lose many, if not all, the world's species of hard corals, along with the diverse reef environments for which they form the foundation. As one hedge against this looming disaster, scientists are freezing coral eggs and sperm to create banks of genetic material. As another, with a $2.1 million grant from the National Oceanic and Atmospheric Administration (NOAA), the Coral Restoration Foundation has undertaken a large-scale coral nursery,[120] growing and propagating staghorn and elkhorn corals at sites scattered around the Florida Keys. The clumps of coral hang from crisscrossed sections of PVC pipe, rather like odd-looking ornaments suspended from a Christmas tree past its prime. Researchers monitor the nursery stock and transplant some of the healthiest specimens to ailing sections of reef.[121]

Countries along the Mesoamerican Barrier Reef have their own coral restoration programs. Belize established 11 coral nurseries in 2009.[122]

The salvation of the Mesoamerican Barrier Reef may even come from a distant relative: Arrecife Alacranes, Scorpion Reef, about 200 miles to the west in the Gulf of Mexico. "Alacranes is very Caribbean-like," explained Nuno Simoes, a marine biologist at the Autonomous National University of Mexico's Yucatán campus in Mérida. "You have a kind of Caribbeanization because of the Yucatán Current, which provides high connectivity."[123]

Protected by the Mexican government as a national park, Alacranes is isolated 66 miles from the Yucatán's northern coast and only reachable by boat. It's also exceptionally large—13 miles across. Except for a small campsite on Isla Perez, the only one of its five islands with any population (a small party of Mexican marines and a lighthouse keeper), the reef lacks tourist accommodations, sparing the creatures that live there from human pressures.

Because many of these species are the same ones that inhabit the Mesoamerican Barrier Reef, they could serve as a "genetic reservoir" in the event that they were wiped out in the Caribbean. Scientists are "just scratching the surface" of Alacranes Reef's potential for conservation of other coral ecosystems, Simoes explained.[124]

"In the Gulf of Mexico, Alacranes is the jewel in the crown," he said. "If Alacranes is protected, it can be used to help recolonize Veracruz, the Florida Keys, and even Bermuda." Not to mention the Mesoamerican Barrier Reef. But there is only so much that recolonization can accomplish. Cautioning against over-optimism, Simoes admitted: "For the

Mexican Caribbean, we may be able to do something, but we will never have pristine reefs."[125]

There isn't a single answer to saving the Mesoamerican Barrier Reef and the other tropical coral reefs around the world. If a solution exists, it has multiple components. We need to understand the causes of the decline of this unique ecosystem, and we need to halt them and to develop and employ ways to ameliorate the damage—at the same time.

"The world is threatened," CORAL's Jennifer Myton said. "The reefs are going into decline. But I've also seen some reefs come back."[126]

We can't wait to understand the threats fully before we address them. But we can work to address them collectively while we take action individually.

19

<hr/>

AND INDIVIDUALLY

What Each of Us Can Do Right Now to Save
the Mesoamerican Barrier Reef

Boyan Slat was still in secondary school when he decided to invent a way to clean up the Great Pacific Garbage Patch. In 2013 at the age of 18, the young Dutchman founded The Ocean Cleanup, a company devoted to wrangling and removing a gyre of plastic twice the size of Texas.[1] Situated midway between Hawaii and California, the gyre circulates clockwise.[2] Toothbrushes, soda straws, cellphone cases—anything imaginable that can be made of plastic and then discarded—becomes trapped in the enormous eddy. The two trillion pieces of trash include tiny fragments from objects degraded by the sun—minute bits that unwary fishes mistake for plankton.[3] They also include "ghost nets;" tossed away by fishermen, these can still snare fishes, birds, and marine mammals.[4]

Slat's concept was ambitious but straightforward: sweep up the plastic trash and haul it to land to be recycled. Crowdfunding his first invention, a 2,000-foot boom trailing a 10-foot long skirt set to capture the debris as it circled by, Slat and his investors soon learned that an idea that had performed well in the development lab didn't work in the open ocean. Waves tore the skirt. The four big anchors designed to help the contraption function like a shoreline failed to slow it sufficiently, with the result that the boom outpaced the gyre. Multiple designs followed. Finally, on October 2, 2019, the company announced that the latest generation of a boom, the System 001/B, worked. Consisting of yellow plastic floats crammed into a long net tube with parachute-style anchors, the new device featured a sweeping function that could be speeded up or slowed down, adjusting its pace to that of the gyre.[5] *New Yorker* writer Carolyn Kormann, who attended one of Slat's 2017 presentations, noted that two years[6] later, backers still had confidence in him.[7]

In 2019 Boyan's team followed with the Interceptor, a barge designed to sweep up plastic debris in rivers before it had a chance to reach the sea. Solar collector panels covered the roof of the 8- × 24-meter vessel, allowing it to operate without polluting. The Ocean Cleanup put Interceptors to work sweeping up trash flushed by the 2019 floods in Jakarta[8] and then in Jamaica and Thailand.[9]

The company went on in 2020 to offer its first commercial product made from the discarded plastic it collects from the Great Pacific Garbage Patch: sunglasses. At 199 euros, they aren't cheap, but they come with their own hard case and a pouch, and all the proceeds go to cleaning up the world's waterways.[10]

Why did Boyan's investors stick with him through multiple unsuccessful iterations of the Garbage Patch sweeping boom? Maybe some ideas for solving environmental problems make too much sense to abandon. And maybe we wish that more individuals were coming forward with potential solutions, and thus we are willing to support those who are—especially if they're young and charismatic and look a bit like Paul McCartney on the Beatles' first tour.

Slat's cleanup technology for the North Pacific Gyre, the largest of the five in the world's oceans, may help remove trash from the Mesoamerican Barrier Reef, as well. A current that feeds off the North Atlantic Gyre sweeps in from the east across South America's Caribbean coast, hitting the reef at the Bay Islands, then traveling north along Guatemala, Belize, and Quintana Roo, finally funneling between the Yucatán and Cuba into the Gulf of Mexico.[11] Plastic from as far away as Africa comes along for the ride.

Although we may not have the vision, charisma, and persistence to do something as ambitious as Boyan Slat's The Ocean Cleanup, we can still act individually to save the Mesoamerican Barrier Reef and the rest of the marine, and terrestrial, environment. Granted, Earth has always been at some point in a climate cycle, becoming warmer or cooler since way before our earliest ancestors emerged from the mire. But this natural process has been gradual, and natural processes act together as a sort of self-correcting system. The difference is that human activity has accelerated the natural cycle dramatically.

Some historians place the dawn of the Industrial Revolution in the last quarter of the eighteenth century, others around 1830. By 1900 it had spread from Great Britain's textile mills and steel plants to factories and utility plants around the world, spewing carbon dioxide, spent fuel and industrial waste into the atmosphere and, either directly or indirectly, into the oceans.[12] What's occurring now is less like the slow natural parade of Ice Ages and balmy eras and more like a sudden cataclysmic event—in ecological terms almost as sudden as the comet or asteroid that crashed off the Yucatán 66 million years ago, wiping out about 75 percent of Earth's species, including the dinosaurs.[13]

As citizens of the world, we need to push collectively for policies that will slow global warming at the same time that we act individually to reduce other negative impacts on Earth's oceans.

The latter is relatively achievable. The first step is simply to be mindful about how we live our daily lives and, to paraphrase eighteenth-century German philosopher Immanuel Kant's Categorical Imperative, act as if we would want everyone to take the same action under similar circumstances. In mundane modern terms, use a refillable aluminum water bottle over a single use plastic bottle not just because your choice will do a tiny amount of good but also because if everyone did that, a lot of plastic would be kept out of the ocean's gyres.

Most of us have become adept at sorting our recyclable trash, and single-use plastic shopping bags have been banned at supermarkets in many parts of the world (although they experienced a comeback during the COVID-19 pandemic when supermarkets forbid their staffs from sacking purchases in customers' reusable bags). Buying products made from recycled materials completes the economic circle. We can also recycle plastic at home. When it comes to keeping leftovers fresh, those yogurt and cottage cheese containers do the job nicely.

Magazine articles, online posts, and even entire books enumerate ways to reduce one's waste generation and carbon footprint without sacrificing standard of living: Buy products in bulk. Reduce the use of petroleum-based household products; for example, remove carpet stains with baking soda and white (not red wine or balsamic) vinegar. Bring a ceramic coffee cup to work. Bundle your errands. Drive an electric, hybrid or at least fuel-efficient vehicle. Use public transportation. If you only require a van to transport visitors or an SUV for occasional weekends camping, rent one when you do and choose a greener way to get around otherwise. When you need to pick up something small or return a couple of DVDs to the library, walk or ride your bike. (Yes, that will burn up calories that will need to be replaced by the food you consume, but you will derive aerobic benefits, and many of us would like to lose a few pounds.)

Eat in season, and choose more from further down the food chain. That doesn't have to mean becoming vegan or even giving up meat. Three ounces of chicken breast tossed with julienned zucchini, carrots, peanut sauce, and soba noodles makes a yummy dinner for two. Served over rice, one pound of dried red beans cooked with one link of andouille sausage and an onion will serve four people as well as six ounces of sirloin and a baked potato per person would. And there'll be plenty left over for lunch.

Save and eat your leftovers. Grow some of your vegetables—or at least your own herbs, which do fine on an apartment balcony, as do cherry tomatoes. If you eat fish, check the Marine Stewardship Council's site "Sustainable Fish to Eat," which allows you to look up your favorite seafood and see how it ranks,[14] or the "Sustainable Seafood Guide" posted by the Monterey Bay Aquarium for a list of species that are abundant and caught or farmed sustainably.[15] The latter comes in a pocket-size format so that you can print it out and carry it to the market—in your reusable shopping bag.

"Be responsible for what you're eating," CORAL's Jennifer Myton advised. "If you're coming here to Roatán and it isn't lobster season, don't eat lobster. And support the local groups that are trying to protect the reefs."[16]

One important step we can take is to reduce our use of sunscreen, especially when we're near tropical reefs, and to avoid altogether sunscreens containing oxybenzone and octinoxate, chemicals known to damage both corals and the marine creatures that share their environment. In a study published in 2015 in the peer-reviewed journal *Archives of Environmental Contamination and Toxicity,* forensic ecotoxicologist Craig Downs found that oxybenzone, a major ingredient in most commercial sunscreens, lowers the temperature at which corals bleach, thus exacerbating the effect of global warming, and damages coral DNA. It also causes coral larvae to form exoskeletons prematurely, encasing themselves in calcium carbonate coffins while still floating in the ocean. In addition, oxybenzone wreaks all sorts of havoc with the reproductive abilities of fishes and mammals.[17]

Lots of things cause bleaching or otherwise damage corals, Downs conceded, but oxybenzone also interferes with their ability to recover.[18]

In 2015 up to 14,000 tons of sunscreen was entering the ocean annually. Not surprisingly, popular tourist beaches received more than their share of this load. Compounding the assault of whatever washes off in the water, oxybenzone and octinoxate are absorbed through the skin, to be excreted later as urine.[19]

In 2018 Hawaii's legislature passed a bill banning the distribution of sunscreen containing either of these two coral-threatening chemicals beginning January 1, 2021. Mexico had already forbidden their use in the country's marine eco-reserves, including popular Xcaret and Xel-Há along the Mesoamerican Barrier Reef.[20]

Inspired by Hawaii, on February 5, 2019, the Key West City Commission tried to enact a similar ban scheduled to go into effect on the same date, voting 6–1 in favor.[21] However, two months later the Florida Senate Community Affairs Committee ruled to prevent cities and counties from prohibiting the sale of those sunscreens, citing doubts about the research findings and concerns about sun safety. Any local government attempting to impose a ban would face a $25,000 fine, and the state's tourism agency would refuse to promote the city. Publix supermarkets, Walmart, and Johnson & Johnson, a major sunscreen manufacturer, had lobbied heavily for the "ban-ban."[22]

But consumers needn't wait for states and cities to ban sunscreen containing oxybenzone and octinoxate. We can buy and use reef safe sunscreen now, both to limit the harm we do at present and to encourage manufacturers to produce and market more. Edgewell Personal Care, maker of Banana Boat and Hawaiian Tropic, also produces a line of sunscreen free of oxybenzone and octinoxate. Smaller corporations, such as RAW Elements USA, also manufacture reef safe sunscreen.[23] It works just fine inland, as well as at the beach. There's evidence that oxybenzone is bad for terrestrial creatures, as well as those in the ocean.

"The ingredients used to make reef-safe sunscreen don't cost any more," explained Kevin Mallory, founder and president of the skincare company YEOUTH (and also my husband's son-in-law), who is considering offering a reef-safe sunscreen online. "The question is demand. If enough people want to buy it, then it will pay manufacturers to make it available."[24]

But even relatively reef safe ingredients, such as zinc and titanium oxide, when blended into products as nanoparticles, can damage corals; many of the other components have yet to be tested for coral toxicity. Using a sunscreen labelled "organic" is no guarantee that it won't harm the reef ecosystem. Lavender, pennyroyal, and other essential oils also work as natural insecticides, so it stands to reason that they might not be good for bugs' aquatic cousins—lobsters, crabs, and shrimps.

One solution is to use less sunscreen, even the reef safe variety, and to use it more wisely. Applying it only to your face, neck, the backs of your hands and the tops of your feet can prevent 90 percent of whatever you might normally smear on from reaching the reef.[25] There's no need to slather sunscreen on your whole body right before you plunge into the sea. When you snorkel, pull on a long-sleeve tee shirt and a pair of board shorts or old cutoffs. If you dive, don't bother lathering up if you'll be sitting under a boat's canopy in your wetsuit; just anoint your face and neck.

We need to protect our skin from ultraviolet rays, but a hat and a long-sleeved shirt, especially one made of UV-blocking materials, will help accomplish that. Lying around sunbathing is a bad idea. No matter what SPF you pour on, that "healthy" bronze glow you seek to develop is your skin's response to a threat at a cellular level, piling on pigment like an umbrella but still experiencing DNA damage that can lead to the development of cancer cells, as well as to premature wrinkles. If you manage to cultivate a "base tan," it will offer about the same level of protection as an SPF 3 sunscreen. People who start out with darker skin have more protection than those with fair complexions, but even sub-Saharan Africans can experience sun damage.[26]

And anyone with any sense should avoid the midday sun. After all, aren't an early morning stroll on the beach and a sunset swim far more pleasant than broiling on the sand at noon?

Granted, when it comes to snorkeling and diving, the colors on the reef are most intense when the sun is highest. Unfortunately, snorkeling and diving also threaten the Mesoamerican Barrier Reef and other coral ecosystems worldwide. A careless kick from a fin or a knock from a dangling instrument console can cause lasting harm to a coral head.

"Not all the guides are receiving training in what are good diving practices," said Marisol Rueda Flores, communications consultant for the Healthy Reefs for Healthy People Initiative (HRI). "Also, some of the guides are not following the rules. For example, they allow people on the whale shark dives to touch the animals." She added that snorkelers in Akumal, where it's very shallow, stand up on the reef, damaging the corals.[27]

If you dive, practice good buoyancy control, and secure your hoses and instruments. If you snorkel, watch where you kick your fins. And when engaging in either activity, never touch anything underwater.

"I have two words of advice," marine biologist Omar Vidal declared. "Show respect. Show respect for people; show respect for nature."[28]

When visiting the Mesoamerican Barrier Reef, one way to show respect for both nature and people is to choose environmentally friendly accommodations. Instead of participating in mass tourism by booking a cruise of the Western Caribbean on a huge ship or by renting a condo in Cozumel, choose an option that gives you an authentic sense of place and allows you to enjoy the natural and cultural environment.

Cruises may seem like a good travel value, but they damage the Mesoamerican Barrier Reef and the other natural wonders worldwide that draw the hordes of tourists who fuel cruise line profits. In 2019, those hordes numbered 30 million. More than 34 percent of them chose the Caribbean. According to the Cruise Lines International Association, whose 50 members ranged from mammoth Carnival to boutique river cruisers, cruising was a $150 billion industry.[29]

In addition to its flagship line, Carnival incorporates Princess, Holland America Lines, and Seabourn, giving it half the world's cruise market.[30] In 2016 the U.S. Justice Department slapped Carnival and Princess with a $40 million fine for discharging oil-polluted waste at sea, presumably because that was cheaper than discharging it legally at port, where it could be processed properly. Part of the penalty was for asking employees to cover up the crime. (Obviously, one of them didn't go along.) Despite assurances from executives that the company would mend its ways, one of their ships was caught again, this time discharging plastic off the Bahamas. The result? Another $20 million in fines.[31]

Even when cruise operators adhere strictly to antipollution laws and go beyond by instituting their own "green" approaches, such as eliminating plastic drinking straws and asking guests to reuse their towels, they imperil the fragile reef environment. Ports tear up fringe reefs to create docks. Lights and engine noise can disrupt spawning; noise pollution can even be fatal to marine creatures. Simply disgorging thousands of passengers, and those crew members lucky enough to have shore leave, creates an ecological and cultural burden on ports of call. And one of those floating behemoths can produce three or four times as much carbon dioxide per passenger as an airplane.[32]

In early 2020, the novel coronavirus set the cruise industry adrift. Responding to late February onboard outbreaks of COVID-19, ports around the world abruptly closed to cruise ships. At the height of the winter season, many of the 350 ships in service wandered from port to port in search of one where their thousands of passengers could disembark. On March 14, the CDC issued a "No Sail" order prohibiting cruise ships from arriving or leaving from U.S. ports for 30 days. On April 9, the agency renewed it for another 100 days. On July 16 the CDC extended the order. When that order expired at the end of September, the CDC extended it again, this time through October 31.[33] In October the agency telegraphed that it would lift the order effective November but could impose restrictions aimed on reducing COVID-19 transmission through November 2021. Declaring that they needed two months to ready their ships for passengers, some cruise companies announced that they would resume sailing in December 2020,[34] but most ended up postponing sailings until late spring 2021 or even canceling their 2021 cruises altogether.[35]

And cruising's giants would set to sea with fewer ships. Deprived of revenues except for some cut-rate advance sales, the industry's leaders struggled to stay afloat. Confronting a loss of $2.9 billion for the quarter ending August 31, 2020, Carnival downsized, jettisoning 13 of its older ships, sending them to shipyards that previously specialized in recycling decommissioned cargo ships. In Aliağa, Turkey, workers removed furniture and gym equipment destined for hotels and corporate offices and set aside light fixtures and decorative items for dealers in maritime memorabilia. Once everything from lifeboats to light bulbs had been stripped, welders would cut up the bulkheads and hulls to sell as scrap steel.[36]

The COVID-19 pandemic sent the entire tourism industry reeling. In the Caribbean, island nations closed, then reopened, then closed again. Some of these countries depended on visitors for 50 to 90 percent of their gross domestic product.[37] The impact on their economies, and on people ranging from stevedores and hotel maids to resort owners, was catastrophic and will only be tallied after this book goes to press. But what about the reefs?

In August 2020 *New York Times* travel writer Nina Burleigh reported that in the waters off Bonaire, divers were spotting more seahorses, a rare sight in recent years. Marine animals in general were coming closer.[38] Ian Drysdale, Honduras country coordinator for HRI, reported a similar phenomenon off Roatán, including an increase in shark sightings.[39]

Comfortable cottage, Isla Marisol Resort, Glover's Reef, Belize, photo courtesy of Isla Marisol Resort.

Granted, Bonaire is way on the other side of the Caribbean from the Mesoamerican Barrier Reef. And, granted, anecdotal reports are no substitute for the rigorous monitoring that scientists will be conducting for years to come. As leisure travel resumes post-pandemic, it may look the same, or almost the same, as it did in 2019, with hordes of budget-conscious travelers booking condos and cruises—and on even larger ships capable of carrying almost 7,000 passengers.

The coronavirus interruption caused communities along the Mesoamerican Barrier Reef to reevaluate their dependence on tourism, especially what Drysdale called the "cash cow" of serving as a port of call for massive cruise ships.

"We need to diversify our production and reduce our economic reliance on tourism," he told me in November 2020. "Our economies are very fragile due to dependency on a single source of funding."[40]

Even as they diversify their economies, the countries along the Mesoamerican Barrier Reef will continue to welcome tourists, but their natural and cultural charms will survive only if visitors invest the thought, and often the money, in mindful travel. Small, eco-friendly resorts tend to cost more than cruises or condos because doing things right takes more staff, and personnel are the biggest expense for any enterprise, but the extra that guests pay above the price of, say, a cruise or a condo in a large complex helps support the health of the reef, as well as the local community.

Those of us who value the natural and human ecosystem will continue to have sustainable choices, whatever our budgets. If you are willing to trade splashy nightlife for rocking in a hammock as you gaze at the stars and sip a bottle of Belikin beer, the price of a palm-shaded cottage on a sandy beach can be competitive. Some small, rustically comfortable compounds even sit right on top of the Mesoamerican Barrier Reef. Take Pelican Beach Resort on South Water Caye, one islet north of the Smithsonian's Carrie Bow Cay Field Station, or Isla Marisol on Glover's Reef. Located 27 miles from the mainland and situated in a Marine Protected Area and UNESCO World Heritage Site, Glover's Reef is as close as it comes to pristine.

Near the fishing village of Hopkins in southern Belize is a true resort that blends quietly elegant accommodations, attentive service and top-notch food with a sincere and persistent commitment to what is called "regenerative travel"—improving, rather than just protecting, the ecosystem, as well as the local community. Hamanasi (Garifuna for a coastal variety of almond tree) belongs to a new generation of eco-resort.

"It's interesting, the whole eco-tourism market," Virginian Dana Krauskopf, who founded and owns Hamanasi in partnership with her husband, David, told me in a 2016 phone interview. "Early on people associated the

The author diving, photo by Ben Phillips.

term with more rustic places, but we've also found that it's necessary to have different standards of accommodation for people."

Admittedly, the nine nights my dive buddy Terry McNearney and I spent at Hamanasi were a splurge. But even if you're staying in a $35 double room (the Winter 2020 price) in the Hopkins Guest House, you can treat yourself to a dinner at Hamanasi, eating by candlelight as you sit on the verandah gazing out toward the cays that mark the edge of the Mesoamerican Barrier Reef. And ordering the lionfish tacos.

ACKNOWLEDGMENTS

As a journalist writing for a lay audience, I have relied on the specialized expertise of the marine biologists, oceanographers, ecologists and other scientists who have responded with unfailing intellectual generosity and patience. As an admirer of the groups dedicated to the well-being of the Mesoamerican Barrier Reef and the people who depend on it, from international organizations to community cooperatives, I am grateful for the perspectives their members have shared. I have cited peer-review papers, print and online publications and books in the text as endnotes and have listed these in the bibliography. In addition, I would like to give special thanks to the individuals who helped me in person or by phone or e-mail.

From Texas A&M University at Galveston, Drs. Ulli Budelmann, Ron I. Eytan, Tom Iliffe, Maria Pia Miglietta, Gilbert Rowe and Anja Schulze were especially helpful early on. Drs. Phil Hastings of the Scripps Institution of Oceanography at the University of California San Diego, Andrea Kealoha of the University of Hawaii Maui College and Nuno Simoes of the Universidad Nacional Autónoma de México's Yucatán campus helped me understand the interconnections in coral reef ecosystems. Dr. Margaret Leshikar-Denton of the Cayman Islands National Museum opened my eyes to the world of underwater archaeology.

Smithsonian Institution coral reef biologist Dr. Nancy Knowlton answered my questions patiently and clearly. The Smithsonian's Dr. Harilaos Lessios described both his personal observations of the lethal long-spined sea urchin epidemic in the early 1980s and his view of the prospects for recovery to date, which Dr. Peter Edwards of California State University, Northridge also helped put into perspective.

I owe a special debt to the Smithsonian Institution's Caribbean Coral Reef Ecosystems (CCRE) Program. Zach Foltz, station manager and dive officer for CCRE, arranged for me to spend a week as a "fly on the wall" at its Carrie Bow Cay Field Station in Belize as Dr. Karen Koltes, John Tschirky and Joanna Walczak conducted longitudinal monitoring of the Mesoamerican Barrier Reef's health. Volunteer station manager Craig Sherwood made me feel at home. Therese Bowman Rath and Dana Krauskopf shared their perspectives as hoteliers seeking to offer ways for visitors to experience the reef ecosystem while respecting it.

Nongovernmental and community-based organizations advocating for the Mesoamerican Barrier Reef and undertaking projects to protect it deserve my appreciation and that of everyone who cares about the environment. I am especially grateful to Jennifer Chapman of Blue Ventures, Stuart Fulton of Comunidad y Biodiversidad (COBI), Gustavo Guerrero Límon of Guardianes del Arrecife, Jennifer Myton of the Coral Reef Alliance (CORAL) and Omar Vidal of the World Wildlife Fund, Mexico. My particular admiration goes out to the Healthy Reefs for Healthy People Initiative (aka Healthy Reefs Initiative or HRI) , which has made a large impact with modest resources and a remarkably efficient team: Dr. Melanie McField, Nicole Katrina Craig, Ian Drysdale, Marisol Rueda Flores, Ana Giró Petersen and Mélina Soto, as well as former HRI team member Roberto Pott.

Thanks to my longtime dive buddy and dear friend Dr. Terry Ann McNearney and the rest of the scuba diving contingent of the Clear Lake Area Ski Club, especially Rick and Laura Heiman and Pam Falk, I have been privileged to visit the Mesoamerican Barrier Reef repeatedly in the best of company.

Marine biologist Dr. Donald Behringer of the University of Florida, zoologist and marine biologist Dr. Richard Brusca of the University of Arizona and coral reef ecologist Dr. Stephen Gittings of the National Oceanic and Atmospheric Administration provided meticulous reviews of my manuscript, often catching potentially embarrassing mistakes. Any remaining errors are mine alone.

Thanks to University Press of Florida, especially Meredith Babb, Sian Hunter, Cindy Durand and Marthe Walters, for recognizing the Mesoamerican Barrier Reef as a multinational ecological treasure and granting me the opportunity to increase public awareness of it, including the challenges it faces.

My agent, Charlotte Raymond, helped shape this book from its inception. I am deeply grateful for her insights, persistence and enthusiasm, for this project and for me as a writer.

For his unflagging love, patience and understanding, I give my heartfelt thanks, as well as my heart, to my husband, Dr. Charles McClelland.

NOTES

Introduction

1. Nancy Knowlton, *Citizens of the Sea: Wondrous Creatures from the Census of Marine Life* (Washington, DC: National Geographic, 2010), 147.

2. Sylvia A. Earle, *The World Is Blue: How Our Fate and the Ocean's Are One.* (Washington, DC: National Geographic, 2009), 10.

3. Knowlton, *Citizens,* 22.

4. Earle, *The World,* 10.

5. National Oceanic and Atmospheric Administration, "How Many Species Live in the Ocean?" Last modified January 7, 2020, accessed January 29, 2020. https://www.oceanservice.noaa.gov/facts/ocean-species.html.

6. Earle, *The World,* 127.

7. Knowlton, *Citizens,* 25.

8. Earle, *The World,* 10.

9. Peter Castro and Michael E. Huber, *Marine Biology: Eleventh Edition.* New York: McGraw Hill Education, 2019, 322–323.

10. Castro and Huber, *Marine Biology,* 326–333.

11. Lucie M. Bland, Tracey J. Regan, Minh Ngoc Dinh, Renata Ferrari, David A. Keith, Rebecca Lester, David Mouillot, Nicholas J. Murray, Houng Anh Nguyen, and Emily Nicholson, "Meso-American Reef: Using Multiple Lines of Evidence to Assess the Risk of Ecosystem Collapse," *Proceedings of the Royal Society B 28,* no. 183. Last modified September 20, 2017, accessed March 10, 2019. https://doi.org/10.1098/rsph.2017.0660.33.

12. At 930 miles long, the New Caledonian Barrier Reef is longer; but rather than lying next to a continent, it circles a huge lagoon and the islands around it and thus differs from both Australia's Great Barrier Reef and the Mesoamerican Barrier Reef. The corals in the Red Sea extend 1,180 miles but did not form as a barrier reef. ("The Longest Coral Reefs in the World." Accessed February 8, 2019. https://www.WorldAtlas.com.)

13. Walter M. Goldberg, "Atolls of the World: Revisiting the Original Checklist," *Atoll Research Bulletin No. 610,* June 28, 2016. Washington, DC: Smithsonian Institution Scholarly Press.

14. Kathryn Hansen, "Reef in the Southern Gulf of Mexico." Posted November 5, 2014, accessed June 20, 2020. https://earthobservatory.nasa.gov/images/85177/reef-in-the-southern-gulf-of-mexico.

15. National Oceanic and Atmospheric Administration, "In What Types of Water Do Corals Live?" Last updated January 7, 2020. http://www.oceanservices.noaa.gov/facts/coralwaters.html.

16. National Oceanic and Atmospheric Administration, "In What Types of Water Do Corals Live?" Last updated January 7, 2020. https://www.oceanservices.noaa.gov/facts/coralwaters.html.

Chapter 1. How to Build a Barrier Reef

1. Charles Darwin, *The Structure and Distribution of Coral Reefs* (Tucson: University of Arizona Press, 1984) (originally published in 1842), 6.

2. Darwin, *Structure and Distribution*, 8.

3. Michael T. Ghiselin, Foreword to Charles Darwin, *The Structure and Distribution of Coral Reefs*, x.

4. Bryce Emley, "How the Diving Bell Opened the Ocean's Depths," *The Atlantic*, March 23, 2017. Accessed May 31, 2020. https://www.theatlantic.com/technology/archive/2017/03/diving-bell/520536/.

5. Siobhan Marie Kilfeather, *Dublin: A Cultural History* (Oxford: Oxford University Press, 2005), 63.

6. Ker Than, "Jacques Cousteau Centennial: What He Did, Why It Matters." Last modified June 11, 2010, accessed September 1, 2017. https://www.news.nationalgeographic.com.

7. Ghiselin, Foreword, xii.

8. Darwin, *Structure and Distribution*, 201.

9. Darwin, *Structure and Distribution*, 196.

10. Thomas Iliffe, in-person interview, January 6, 2016.

11. Victor J. Polyak, Bogdan P. Onac, Joan J. Fornós, Carling Hay, Yemane Asmerom, Jeffrey A. Dorale, Joaquín Ginés, Paola Tuccimei, and Angel Ginés. "A Highly Resolved Record of Relative Sea Level in the Western Mediterranean Sea During the Last Interglacial Period," *Nature Geoscience* 11, 860–864, 2018. https://doi.org/10.1038/s41561-018-0222-5.

12. Jorgen Olsen, Kenneth Meland, Henrik Glenner, Peter J. Van Hergstrum, and Thomas M. Iliffe, "Xibalbanus Cozumelensis, a New Species of Remipeda (Crustacea) from Cozumel, Mexico, and a Molecular Phylogeny of Xibalbanus on the Yucatán Peninsula," *European Journal of Taxonomy*, no. 316, 1. Posted May 31, 2017, accessed May 31, 2020. https://europeanjournaloftaxonomy.eu/index.php/ejt/article/view/441.

13. Iliffe, in-person interview, January 6, 2016.

14. González, Arturo H., Carmen Rojas Sandoval, Eugenio Acevez Núñez, Jerónimo Avilés Olguín, Santiago Analco Ramírez, Octavio del Río Lara, Pilar Luna

Erreguerena, Adriana Velázquez Morlet, Wolfgang Stinnesbeck, Alejandro Terrazas Mata, and Martha Benavente Sanvicente, "Evidence of Early Inhabitants in Submerged Caves in Yucatán, Mexico," in *Underwater and Maritime Archeology in Latin America and the Caribbean,* eds. Margaret Leshikar-Denton and Pilar Luna Erreguerena (Walnut Creek, CA: Left Coast Press, 2008), 137.

15. González González et al., "Evidence," 138.

16. González González et al., "Evidence," 137–138.

17. González González et al., "Evidence," 139.

18. Rolex.org, "Arturo González: In Search of the First Americans." Posted 2008, accessed January 18, 2020. http://rolex.org/eng/rolex-awards/exploration/arturo-gonzalez.

19. González González et al., "Evidence," 137.

20. González González et al., "Evidence," 127.

21. Edward H. Thompson, "Atlantis Not a Myth," *Alpena Weekly Argus,* February 4, 1880.

22. Carmen Rojas Sandoval, Arturo H. González González, Alejandro Terrazas Mata, and Martha Benavente Sanvicente, "Mayan Mortuary Deposits in the Cenotes of Yucatán and Quintana Roo, Mexico," in *Underwater and Maritime Archeology in Latin America and the Caribbean,* eds. Margaret Leshikar-Denton and Pilar Luna Erreguerena (Walnut Creek, CA: Left Coast Press, 2008), 144.

23. *New World Encyclopedia,* "Edward Herbert Thompson." Last modified September 22, 2017, accessed June 21, 2020. https://www.newworldencyclopedia.org/entry/Edward_Herbert_Thompson.

24. Rojas Sandoval et al., "Mayan," 144.

25. Rojas Sandoval et al., "Mayan," 146.

26. González González et al., "Evidence," 138.

27. González González et al., "Evidence," 141.

28. Eugene A. Shinn, R. B. Halley, J. H. Hudson, B, Lidz, D. M. Robbin, and I. G. Macintyre, "Geology and Sediment Accumulation Rates at Carrie Bow Cay, Belize," in Smithsonian Contributions to the Marine Sciences, no. 12, *The Atlantic Barrier Reef Ecosystem at Carrie Bow Cay, Belize, I: Structure and Communities,* eds. Klaus Rützler and Ian G. Macintyre (Washington, DC: Smithsonian Institution Press, 1982), 63.

29. Shinn et al., "Geology and Sediment," 72.

30. Thomas M. Iliffe, "Anchialine Caves and Cave Fauna of the World," *National Geographic.* Last modified March 2010, accessed August 30, 2017. http://magma.nationalgeographic.com/ngm/0310/feature4/index.html.

31. Iliffe, "Anchialine Caves."

32. Eden Garcia and Karie Holterman, "Calabash Caye, Turneffe Islands Atoll, Belize," in *CARICOMP Caribbean Coral Reef, Seagrass and Mangrove Sites,* vol. 3, ed. Björn Kierfve, (Paris: UNESCO, 1998), 67.

33. Garcia, "Calabash," 71.

34. Garcia, "Calabash," 69.

35. Gilbert Rowe, in-person interview, January 7, 2016.

36. Nuno Simoes, phone interview, January 21, 2020.

37. Simoes, phone interview, January 21, 2020.

38. Rowe, in-person interview, January 7, 2016.

39. Rowe, in-person interview, January 7, 2016.

40. Rowe, in-person interview, January 7, 2016.

41. Elizabeth Kolbert, "Unnatural Selection: What Will It Take to Save the World's Reefs and Forests?" *New Yorker*, April 18, 2016, 25.

42. Darwin, *Structure and Distribution*, 15.

43. Earle, *The World*, 54.

44. Rowe, in-person interview, January 7, 2016.

45. Charles Birkeland, "Introduction," in *Life and Death of Coral Reefs*, ed. Charles Birkeland (New York: Chapman & Hall, 1997). 8.

46. Peter Castro and Michael E. Huber, *Marine Biology: Eleventh Edition* (New York: McGraw Hill Education, 2019), 329.

47. Iliffe, in-person interview, January 6, 2016.

48. Daniel M. Alongi, "Carbon Cycling and Storage in Mangrove Forests," *Annual Review of Marine Science 6*, January 2014, 195. Accessed May 31, 2020. https://doi.org/10.1146/annureviewmarinne-010213-B5020.

49. Melanie McField, *Report Card for the Mesoamerican Reef: An Evaluation of Ecosystem Health 2010* (Playa del Carmen, Mexico: Healthy Reefs for Healthy People, 2010), 14.

50. Rowe, in-person interview, January 7, 2016.

51. National Oceanic and Atmospheric Administration, "How Does Sand Form?" Last modified April 9, 2020, accessed May 30, 2020. https://www.oceanservice.noaa.gov/facts/sand.html.

52. Simon R. Therrold and Jonathan A. Hare, "Otolith Applications in Reef Fish Ecology," in *Coral Reef Fishes*, ed. Peter F. Sale (San Diego: Academic Press, 2002), 262.

53. Eugene H. Kaplan, *Coral Reefs: A Guide to the Common Invertebrates and Fishes of Bermuda, the Bahamas, Southern Florida, the West Indies, and the Coast of Central and South America* (Norwalk, CT: The Easton Press, 1982), 146.

54. "Animals," *National Geographic*. Accessed July 3, 2019. *https://www.national-geographic.com/animals/invertebrates/group/nudibranchs/*.

55. Kaplan, *Coral Reefs*, 146.

56. Peter J. Etnoyer et al., "Exploration and Mapping of the Deep Mesoamerican Reef," *Oceanography 28*, no. 1, March 2015.

57. Klaus Rützler and Ian G. Macintyre, "The Habitat Distribution and Community Structure of the Barrier Reef Complex at Carrie Bow Cay, Belize," in Smithsonian Contributions to the Marine Sciences, no 12, *The Atlantic Barrier Reef Ecosystem at Carrie Bow Cay, Belize, I: Structure and Communities*, eds. Klaus Rützler and Ian G. Macintyre (Washington, DC: Smithsonian Institution Press, 1982), 11.

58. Rützler and Macintyre, "Habitat," 9.

59. Castro and Huber, *Marine Biology,* 329.

60. Rützler and Macintyre, "Habitat," 6.

61. Erika Gress, Joshua D. Voss, Ryan J. Eckert, Gwilym Rowlands, and Dominic A. Andradi-Brown, "The Mesoamerica Reef," in *Mesophotic Coral Ecosystems,* eds. Yossi Loya, Kimberley A. Puglise, and Tom C. L. Bridge (Cham, Switzerland: Springer Nature Switzerland AG, 2019), 73.

62. Rützler and Macintyre, "Habitat," 37.

63. Yossi Loya, Kimberley A. Puglise, and Tom C. L. Bridge, eds., *Mesophotic Coral Ecosystems* (Cham, Switzerland: Springer Nature Switzerland AG, 2019), ix.

64. Loya, Puglise, and Bridge, *Mesophotic,* ix.

65. Gress, Voss, Eckert, Rowlands, and Andradi-Brown, "The Mesoamerican," 71.

66. Gress, Voss, Eckert, Rowlands, and Andradi-Brown, "The Mesoamerican," 78.

67. Gress, Voss, Eckert, Rowlands, and Andradi-Brown, "The Mesoamerican," 79.

68. Oliver Milman, "Below Bermuda, the Quest Begins to Map the Damage We Are Doing to the Deep Sea," *The Guardian.*

69. McField, *Report Card 2010,* 7.

70. Sarah Egner, "Differentiating Coral Bleaching and Coral Mortality: A Case Study from the Great Barrier Reef," *Alert Diver 33,* no. 1 (Winter 2017), 102–105.

71. Castro and Huber, *Marine Biology,* 319.

72. Warren E. Burgess, *Corals* (Neptune, NJ: T.F.H. Publications, 1979), 9.

73. Kolbert, "Unnatural Selection," 25.

74. Knowlton, *Citizens,* 116.

75. Les Kaufman and John Tschirky, *Living with the Sea: Local Efforts Buffer Effects of Global Change* (Arlington, VA: Conservation International, Science and Knowledge Division, 2010), 5.

76. Kaufman and Tschirky, *Living,* 16.

77. Melanie McField, Patricia Kramer, Ana Giró Petersen, Mélina Soto, Ian Drysdale, Nicole Craig, and Marisol Rueda Flores, *Mesoamerican Reef Report Card 2020* (Playa del Carmen, Mexico: Healthy Reefs for Healthy People), 6. Posted February 2020, accessed June 16, 2020. https://www.healthyreefs.org/cms/wp-content/uploads/2020/02/2020-Report-Card-MAR.pdf.

78. McField et al., *Mesoamerican Reef,* 6.

79. McField et al., *Mesoamerican Reef,* 10–11.

80. McField et al., *Mesoamerican Reef,* 10–11.

81. Intergovernmental Panel on Climate Change, press release, September 24, 2019. http://www.ipcc.ch/2019/09/25/srocc-press-release.

82. Knowlton, *Citizens,* 195.

83. Earle, *The World,* 258.

84. Earle, *The World,* 170.

85. Knowlton, *Citizens,* 195.

86. Knowlton, *Citizens,* 195.

Chapter 2. Who Knew, and When? Discovering the Caribbean's Greatest Barrier Reef

1. Klaus Rützler and Ian G. Macintyre, "Preface," in Smithsonian Contributions to the Marine Sciences, no. 12, *The Atlantic Barrier Reef Ecosystem at Carrie Bow Cay, Belize, I: Structure and Communities*, eds. Klaus Rützler and Ian G, Macintyre (Washington, DC: Smithsonian Institution Press, 1982), xiii.

2. Smithsonian Institution, "About." Accessed October 16, 2016, https:/www.si.edu/about/organization.

3. Smithsonian, "About."

4. Rützler and Macintyre, "Habitat," 9.

5. Jorge Cortés, Hazel A. Oxenford, Brigitta I. van Tussenbroek, Eric Jordán-Dahlgren, Aldo Cróquer, Carolina Bastidas, and John C. Ogden, "The CARICOMP Network of Caribbean Marine Laboratories (1985–2007): History, Key Findings, and Lessons Learned," *Frontiers in Marine Science*, January 14, 2019. Accessed June 1, 2020. https://doi.org/10.3389/frmqrs.2018.00519.

6. Rützler and Macintyre, "Preface," i–vi.

7. Therese Bowman Rath, phone interview, October 11, 2016.

8. Rützler and Macintyre, "Preface," i–vi.

9. World Population Review. "Belize Population 2021 (Live)." Posted February 2021, accessed February 12, 2021. http://worldpopulationreview.com/countries/belize-population.

10. Belize.com. Accessed March 12, 2019. https://belize.com/demographics/.

11. Belize.com. Accessed March 12, 2019. https://belize.com/history-of-the-garufina-people.

12. Belize.com. Accessed March 12, 2019. https://belize.com/demographics/.

13. John Tschirky, in-person interview, September 12, 2016.

14. World Wildlife Fund, "Under the Sea." Posted January 2020, accessed June 1, 2020. *https://www.panda.org/knowledge_hub/where_we_work/mesoamerican_reef.*

15. James N. Norris and Katina E. Bucher, "Marine Algae and Seagrass from Carrie Bow Cay, Belize," in Smithsonian Contributions to the Marine Sciences, no. 12, *The Atlantic Barrier Reef Ecosystem at Carrie Bow Cay, Belize, I: Structure and Communities*, eds. Klaus Rützler and Ian G. Macintyre (Washington, DC: Smithsonian Institution Press, 1982), 167.

16. World Wildlife Fund, "Under the Sea."

17. Lucie M. Bland, Tracey J. Regan, Minh Ngoc Dinh, Renata Ferrari, David A. Keith, Rebecca Lester, David Mouillot, Nicholas J. Murray, Houng Anh Nguyen, and Emily Nicholson, "Meso-American Reef: Using Multiple Lines of Evidence to Assess the Risk of Ecosystem Collapse," *Proceedings of the Royal Society B 28*, no. 183. Last modified September 20, 2017, accessed March 10, 2019. https://dol.org/10.1098/rsph.2017.0660.

18. Bland et al., "Mesoamerican."

19. Bland et al., "Mesoamerican."

20. Bland et al., "Mesoamerican."

21. Rützler and Macintyre, "Habitat," 1.

22. Belize Audubon Society, "Half Moon Caye Natural Monument." Last modified 2014–2016, accessed March 12, 2019. https://www.belizeaudobon.org.

23. World Wildlife Fund, "Mesoamerican Barrier Reef." Last modified April 11, 2012, accessed March 9, 2019. https://www.worldwildlife.org/places/meso american-reef.

24. Melanie McField, *Mesoamerican Reef: An Evaluation of Ecosystem Health, Healthy Reefs for Healthy People, 2015 Repot Card,* 8. Accessed March 13, 2019. https://www/healthyreefs.org/cms/wp-content/uploads/2015/05/MAR-EN-small.pdf.

25. McField, *Mesoamerican 2015,* 8.

26. Stuart Fulton, phone interview, January 20, 2020.

27. Julian Smith, "Bracing for Impact," *Nature Conservancy,* Winter 2018, 50–59.

28. Gustavo Guerrero, phone interview, January 15, 2020.

29. Melanie McField, phone interview, October 10, 2016.

30. Tschirky, in-person interview, September 12, 2016.

31. McField, phone interview, October 10, 2016.

32. Ric Hajovsky, "How Cozumel's Tourism Began." Last modified 2011, accessed March 14, 2019. http://everythingcozumel.com/cozumel-history/cozumels-tourism-industry-began/.

33. Michael Carlton, "Ixtapa: Mexico's Gem of a Resort," *Spokesman-Review,* January 24, 1982, B6.

34. Jules Siegel, *Cancún User's Guide,* 204. Last modified 2006. Lulu.com.

35. Knoema, "Quintana Roo: Total Population." Posted January 2021, accessed February 14, 2021. http://knoema.com/atlas/Mexico/Quintana-Roo/topics/Demo graphics/Key-Indicators/Population.

36. Pro México Trade and Investment. Accessed March 12, 2019. http://mim.pro-mexico.gob.mx/work/models/mim/ . . . /PDF/FE-QUINTANA_ROO.

37. Sherri Davis, "Weekly Cruise Ship Arrival and Departure Information," *Cozumel Insider.* Last modified 2019, accessed March 15, 2019. https://cozumelinsider.com/?Page=CruiseShips.

38. Therese Bowman Rath, phone interview, October 11, 2016.

Chapter 3. Uneasy Symbiosis: The Relationship between the Mainland and the Reef

1. McField, *Mesoamerican 2010,* 2.

2. McField, *Mesoamerican 2010,* 2.

3. Lee Kaufman and John Tschirky, *Living with the Sea: Local Efforts Buffer Effects of Global Change* (Arlington, VA: Conservation International, Science and Knowledge Division, 2010), 10.

4. Kaufman and Tschirky, *Living,* 11.

5. World Wildlife Fund, "Our Work: Biodiversity." Posted 2019, accessed February 1, 2020. https://www.wwf.panda.org/our-work/biodiversity/biodiversity.

6. Knowlton, *Citizens*, 188.

7. Knowlton, *Citizens*, 181.

8. Björn Kievfre, ed., *CARICOMP Caribbean Coral Reef, Seagrass and Mangrove Sites*, vol. 3, (Paris: UNESCO, 1998), 5.

9. Gustavo Guerrero, phone interview, January 15, 2020.

10. Eden Garcia and Karie Holterman, "Calabash Caye, Turneffe Islands Atoll, Belize," *CARICOMP Caribbean Coral Reef, Seagrass and Mangrove Sites*, ed. Björn Kierfve, vol. 3, 67–78 (Paris: UNESCO, 1998), 73.

11. Kenneth Brower, "Meso Amazing," *National Geographic*, 431. Last modified October 2012, accessed March 9, 2019. https://www.nationalgoegraphic.com/magazine/2012.10/mesoamerican-reef.

12. Gilbert Rowe, in-person interview, January 7, 2016.

13. Kennedy Warne, "Forests of the Tide (Mangroves)," *National Geographic 211*, no. 2, February 2007, 132–151.

14. Brower, "Meso Amazing," 431.

15. Knowlton, *Citizens*, 177.

16. Tschirky, in-person interview, September 8, 2016.

17. Smithsonian Tropical Monitoring Networks, "Carrie Bow Cay." Posted 2020, accessed June 23, 2020. https://www.nmnhmp.riocean.com.

18. Tom Iliffe, in-person interview, January 6, 2016.

19. McField, *Mesoamerican 2010*, 14.

20. Kievfre, *CARICOMP*, 5.

21. Castro and Huber, *Marine Biology*, 112.

22. John C. Ogden, "Book Review: Seagrasses, Ecology and Conservation," *Marine Ecology 27*, no. 4, December 2006 (The Netherlands: Springer, 2006), 431.

23. Ogden, "Book Review," 431.

24. Miguel A Mateo., Just Cebrián, Kenneth Dunton, and Troy Mutchler, "Carbon Flux in Seagrass Ecosystems," in *Seagrasses: Biology, Ecology and Conservation*, eds. A.W.D. Larkum, R. J. Orth, and C. M. Duarte (The Netherlands: Springer, 2006), 161.

25. Brower, "Meso Amazing," 431.

26. Tschirky, in-person interview, September 8, 2016.

27. Tschirky, in-person interview, September 8, 2016.

28. Marisol Rueda Flores, phone interview, November 3, 2016.

29. Mateo et al., "Carbon," 160.

30. Mateo et al., "Carbon," 186.

31. Mateo et al., "Carbon," 159.

32. Ogden, "Book," 432.

33. National Oceanic and Atmospheric Administration, "Queen Conch." Accessed June 15, 2020. https://www.cio.noaa.gov/service_programs/prplans/pdfs/ID236_Queen_Conch_Final_Status_Report.pdf.

34. Knowlton, *Citizens*, 188.

35. Kaufman and Tschirky, *Living*, 6.
36. Castro and Huber, *Marine Biology*, 318.
37. Kaplan, *Coral Reefs*, 78.
38. Kaplan, *Coral Reefs*, 78.
39. Kaplan, *Coral Reefs*, 78.
40. Knowlton, *Citizens*, 140.
41. Nicole D. Fogarty and Kristen L. Marhaver, "Coral Spawning, Unsynchronized: Breakdown in Coral Spawning May Threaten Coral Reef Recovery," *Science 365*, no. 6457, 2019, 987.
42. Fogarty and Marhaver, "Coral Spawning," 988.
43. Fogarty and Marhaver, "Coral Spawning," 987.
44. Kaplan, *Coral Reefs*, 80.
45. Kaplan, *Coral Reefs*, 81.
46. Kaplan, *Coral Reefs*, 6.
47. Kaplan, *Coral Reefs*, 6.
48. Mary W. Rice and Ian G. Macintyre, "Distribution of Sipuncula in the Coral Reef Community, Carrie Bow Cay, Belize," in Smithsonian Contributions to the Marine Sciences, no. 12, *The Atlantic Barrier Reef Ecosystem at Carrie Bow Cay, Belize, I: Structure and Communities*, eds. Klaus Rützler and Ian G. Macintyre (Washington, DC: Smithsonian Institution Press, 1982), 318.
49. Kaplan, *Coral Reefs*, 6.
50. Kaplan, *Coral Reefs*, 8.
51. Kaplan, *Coral Reefs*, 7.
52. Knowlton, *Citizens*, 110.

Chapter 4. Masters of the Aquatic Commute: Sea Turtles

1. Amie Braütigam and Karen L. Eckert, *Turning the Tide: Exploitation, Trade and Management of Marine Turtles in the Lesser Antilles, Central America, Colombia and Venezuela*, (Cambridge, UK: TRAFFIC International and CITES Secretariat, 2006), 518.
2. Braütigam and Eckert, *Turning*, 9.
3. Knowlton, *Citizens*, 50.
4. Knowlton, *Citizens*, 67.
5. Knowlton, *Citizens*, 50.
6. Braütigam and Eckert, *Turning*, 9.
7. Melissa Gaskill, "Watching Wildlife from Space," *Alert Diver 35*, no. 1, Winter 2019, 20.
8. Braütigam and Eckert, *Turning*, 336.
9. National Oceanic and Atmospheric Administration, "Sea Turtles." Accessed February 18, 2019. https://www.fisheries.noaa.gov/sea-turtles.
10. Braütigam and Eckert, *Turning*, 519–520.
11. Braütigam and Eckert, *Turning*, 3.
12. National Wildlife Federation, "Sea Turtle: Hawksbill." Accessed February 14,

2021. http://nwf.org/Educational-Resources/Wildlife-Guide/Reptiles/Sea-Turtles/Hawksbill-Sea-Turtle.

13. Braütigam and Eckert, *Turning*, 3.

14. Braütigam and Eckert, *Turning*, 521–522.

15. Braütigam and Eckert, *Turning*, 520.

16. Sea Turtle Conservancy, "Information about Sea Turtles: Frequently Asked Questions." Accessed February 19, 2019. http://conserveturtles.org/information-sea-turtles.

17. *National Geographic*, "Sea Turtles Match Breathing to Diving Depths?" Last modified November 23, 2010, accessed February 19, 2019. https://www.nationalgeographic.com/news/2010/11/10123-leatherback-sea-turtle-buoyancy-vin-video/.

18. Braütigam and Eckert, *Turning*, 11.

19. Braütigam and Eckert, *Turning*, 12.

20. Braütigam and Eckert, *Turning*, 11.

21. Braütigam and Eckert, *Turning*, 5.

22. Sea Turtle Conservancy, "Information."

23. Michael R. Heithaus, Tessa Alcoverro, Rohan Arthur, Derek A. Burkholder, Kathryn A. Coates, Marijolijn A. Christiansen, Nachiket Kelkar, Sarah A. Manuel, Aron J. Wirsing, W. Judson Kenworthy, and James W. Fourqurean, "Seagrasses in the Age of Sea Turtle Conservation and Shark Overfishing," *Frontiers in Marine Science*, August 5, 2014. Accessed June 3, 2020. https://doi.org/10.3389/fmars.2014.00028.

24. National Oceanic and Atmospheric Administration, "Sea Turtles."

25. National Oceanic and Atmospheric Administration, "Sea Turtles."

26. National Oceanic and Atmospheric Administration, "Sea Turtles."

27. Share the Beach, "Alabama Sea Turtle Nesting Facts." Accessed February 25, 2019. https://www.alabamaseaturtles.com/nesting-sea-turtle-statistics.

28. National Oceanic and Atmospheric Administration, "Sea Turtles."

29. Braütigam and Eckert, *Turning*, 6.

30. Braütigam and Eckert, *Turning*, 5.

31. Edward A. Standora and James R. Spotila, "Temperature Dependent Sex Determination in Sea Turtles," *Copeia 1985*, no. 3, August 1985, 711.

32. Standora and Spotila, "Temperature," 711.

33. Knowlton, *Citizens*, 156.

34. Sea World Parks and Entertainment, "Sea Turtles." Accessed February 26, 2019. https://www/seaworld.org/animals/all-about/sea-turtles/characteristics.

35. Richard D. Reina, T. Todd Jones, and James R. Spotik, "Salt and Water Regulation by the Leatherback Sea Turtle *Dermochelys coracea*," *Journal of Experimental Biology*, no. 205, 2002, 1853–1860, The Company of Biologists, Ltd., Great Britain. Accepted April 4, 2002, accessed June 4, 2020. https://jeb.biologists.org/content/jexbio/205/13/1853/full.pdf.

36. Sportalsub.net, "Danish Stig Severinsen Sets New Guinness World Record for Dynamic Apnea at Sea in Mexico." Posted December 24, 2020, accessed February 16, 2021. http://www.sportalsub.net/en/stig-severinsen-record-guiness-2020/.

37. Knowlton, *Citizens*, 61.

38. Sea Turtle Conservancy, "Information."

39. Earle, *The World*, 207.

40. Knowlton, *Citizens*, 199.

41. Braütigam and Eckert, *Turning*, 335.

42. Knowlton, *Citizens*, 199.

43. Braütigam and Eckert, *Turning*, 344.

44. Sea World Parks and Entertainment, "All about Sea Turtles." Accessed February 18, 2019. https://www/seaworld.org/animals/all-about/sea-turtles.

45. Sea Turtle Conservancy, "Information about Sea Turtles: Threats to Sea Turtles." Accessed February 22, 2019. http://conserveturtles.org/information-sea-turtle-threats-sea-turtles.

46. National Oceanic and Atmospheric Administration, "Sea Turtles."

47. National Oceanic and Atmospheric Administration, "Sea Turtles."

48. Beth Brost, Blair Witherington, Anne Meylan, Erin Leone, Llewellyn Ehrhart, and Dean Bagley, "Sea Turtle Hatching Production from Florida (USA) Beaches, 2002–2012, with Recommendations for Analyzing Hatching Success," *Endangered Species Research*, 2015, 2753–2768.

49. According to its website, https://www.cites.org, accessed February 23, 2019, "CITES (the Convention on International Trade in Endangered Species of Wild Fauna and Flora) is an international agreement between governments adopted in 1963 at a meeting of members of IUCN (The World Conservation Union). Its aim is to ensure that international trade in specimens of wild animals and plants does not threaten their survival. CITES is a voluntary international agreement that does not take the place of a country's national laws."

50. Braütigam and Eckert, *Turning*, 18.

51. Braütigam and Eckert, *Turning*, 23.

52. Braütigam and Eckert, *Turning*, 335.

53. Braütigam and Eckert, *Turning*, 336.

54. Braütigam and Eckert, *Turning*, 343.

55. Braütigam and Eckert, *Turning*, 349.

56. Braütigam and Eckert, *Turning*, 344.

57. Braütigam and Eckert, *Turning*, 23.

58. Braütigam and Eckert, *Turning*, 12.

59. Braütigam and Eckert, *Turning*, iv.

60. Braütigam and Eckert, *Turning*, vii.

61. Braütigam and Eckert, *Turning*, iv.

62. Braütigam and Eckert, *Turning*, 18.

63. Braütigam and Eckert, *Turning*, 399.

64. Braütigam and Eckert, *Turning*, 399

65. Braütigam and Eckert, *Turning*, 406.

66. Braütigam and Eckert, *Turning*, 406.

67. Central Intelligence Agency, "Poverty in Guatemala." Accessed February 24,

2019. https://www.cia.gov/library/publications/the-world-facebook/geos/print-gt.html.

68. Braütigam and Eckerd, *Turning*, 382.

69. Braütigam and Eckerd, *Turning*, 386.

70. Braütigam and Eckerd, *Turning*, viii.

71. Braütigam and Eckerd, *Turning*, 25.

72. Braütigam and Eckerd, *Turning*, 392.

73. Braütigam and Eckerd, *Turning*, 392.

74. Emily Petsko, "Tackling a Triple Threat: Belize Banned Bottom Trawling, Off-shore Drilling, and Now Gillnets," Oceana, December 8, 2020. http://Oceana.org/blog/tackling-a-triple-threat-belize-banned-bottom-trawling-offshore-drilling-and-now-gillnets.

75. Braütigam and Eckerd, *Turning*, 344.

76. Ohtadmin, "Government Moves to Control Gill Net Use in Belizean Waters," *The New York Carib News*, December 18, 2019.

77. Ohtadmin, "Government."

78. Tschirky, in-person interview, September 12, 2016.

79. Braütigam and Eckerd, *Turning*, 336.

80. Braütigam and Eckerd, *Turning*, 336.

81. Knowlton, *Citizens*, 199.

Chapter 5. The Life Translucent: Jellyfishes and Their Kin

1. Castro and Huber, *Marine Biology*, 121.

2. Kaplan, *Coral Reefs*, 62.

3. Kaplan, *Coral Reefs*, 62–63.

4. Knowlton, *Citizens*, 162.

5. North Carolina Aquarium at Fort Fisher, "Inside a Sea Turtle's Mouth." Posted February 9, 2017, accessed June 24, 2020. https://seaturtlexploration.com/inside-of-a-sea-turtles-mouth/.

6. Juli Berwald, *Spineless: The Science of Jellyfish and the Art of Growing a Backbone* (New York: Riverhead Books, 2017), 172.

7. Miglietta, in-person interview, January 6, 2016.

8. Berwald, *Spineless*, 50.

9. Kaplan, *Coral Reefs*, 55.

10. Ronald J. Larson, "Medusae (Cnidaria) from Carrie Bow Cay, Belize," in Smithsonian Contributions to the Marine Sciences, no. 12, *The Atlantic Barrier Reef Ecosystem at Carrie Bow Cay, Belize, I: Structure and Communities*, eds. Klaus Rützler and Ian G. Macintyre (Washington, DC: Smithsonian Institution Press, 1982), 353.

11. E. Suarez-Morales, L. Segura-Puertes, and R. Gresca, "A Survey of the Reef-Related Medusa (Cnidaria) Community in the Western Caribbean," *Gulf Research Reports 11*, no. 1, January 1999.

12. Barry W. Spracklin, "Hydroidea (Cnidaria: Hydroza) from Carrie Bow Cay, Belize," in Smithsonian Contributions to the Marine Sciences, no. 12, *The Atlantic Barrier*

Reef Ecosystem at Carrie Bow Cay, Belize, I: Structure and Communities, eds. Klaus Rüt-zler and Ian G. Macintyre (Washington, DC: Smithsonian Institution Press, 1982), 239.

13. Kaplan, *Coral Reefs,* 65.

14. Kaplan, *Coral Reefs,* 65.

15. Kaplan, *Coral Reefs,* 57.

16. Kaplan, *Coral Reefs,* 57.

17. Berwald, *Spineless,* 55.

18. Kaplan, *Coral Reefs,* 57

19. Maria Pia Miglietta, "Genomics of the 'Immortal' Jellyfish *Turritopsis dohrinii*," Texas A&M Galveston website. Posted 2020, accessed June 24, 2020. https://www.tamug.edu/miglietta/Research.html.

20. Maria Pia Miglietta, in-person interview, January 6, 2016.

21. Miglietta, in-person interview, January 6, 2016.

22. Berwald, *Spineless,* 155.

23. *The Guardian,* "Nobel Prize in Physiology or Medicine 2012: As It Happened," U.S. Edition online. Last modified October 8, 2012, accessed January 1, 2019, https://www.theguardian.com/sciene/blog/2012/oct/08.

24. Miglietta, in-person interview, January 6, 2016.

25. Berwald, *Spineless,* 72.

26. Berwald, *Spineless,* 66.

27. Berwald, *Spineless,* 13.

28. Knowlton, *Citizens,* 9.

29. Knowlton, *Citizens,* 67.

30. Kaplan, *Coral Reefs,* 66.

31. Knowlton, *Citizens,* 39.

32. Berwald, *Spineless,* 122.

33. Berwald, *Spineless,* 72.

34. Berwald, *Spineless,* 127–128.

35. Berwald, *Spineless,* 125.

36. Berwald, *Spineless,* 9.

37. Richard D. Reina, T. Todd Jones, and James R. Spotik, "Salt and Water Reg-ulation by the Leatherback Sea Turtle *Dermochelys coracea*," *Journal of Experimental Biology,* no. 205, 2002, 1853, The Company of Biologists, Ltd., Great Britain. Accepted April 4, 2002, accessed June 4, 2020. https://jeb.biologists.org/content/jexbio/205/13/1853/full.pdf.

38. Jennifer E. Purcell, Shin-ichi Uye, and Wen-Tseng Lo, "Anthropogenic Causes of Jellyfish Blooms and Their Direct Consequences for Humans: A Review," *Marine Ecology Progress Series 350,* 154. Published November 4, 2007, accessed June 4, 2020. https://int-res.com/articles/meps2007/350/m350p153.pdf.

39. Berwald, *Spineless,* 171.

40. Miglietta, in-person interview, January 6, 2016.

41. Miglietta, in-person interview, January 6, 2016.

42. Berwald, *Spineless,* 171.

43. Ted Olson, "Invasion of the Jellyfish," *Scholastic Action.* February 2018 online edition, last modified February 2018, accessed January 13, 2019. https://action.scholastic.com.

44. Miglietta, in-person interview, January 6, 2016.

45. Berwald, *Spineless,* 26.

46. Berwald, *Spineless,* 26.

47. Berwald, *Spineless,* 25.

48. Berwald, *Spineless,* 24.

49. Olson, "Invasion."

50. Berwald, *Spineless,* 38.

51. Purcell et al., "Anthropogenic," 153.

Chapter 6. Hanging on for Dear Life: Creatures Anchored to the Reef

1. Thomas Bulfinch, *Bulfinch's Mythology* (London, Spring Books, 1964), 85.

2. Jerry and Idaz Greenberg, *The Living Reef,* (Miami: Seahawk Press, 1987), 30.

3. Myriam Lacharité and Anna Metaxas, "Early History of Deep-Water Gorgonian Corals May Limit Their Abundance," *PLoS ONE.* Posted June 10, 2013, accessed June 25, 2020. https://doi.org/10.1371/journal.pone.0065394.

4. Katherine Muzik, "Octocorallis (Cnidaria) from Carrie Bow Cay, Belize" in Smithsonian Contributions to the Marine Sciences, no. 12, *The Atlantic Barrier Reef Ecosystem at Carrie Bow Cay, Belize, I: Structure and Communities,* eds. Klaus Rützler and Ian G. Macintyre (Washington, DC: Smithsonian Institution Press, 1982), 305.

5. Muzik, "Octocorallis," 303.

6. Greenberg, *Living Reef,* 25.

7. Burgess, *Corals,* 9.

8. Robert Kinzie III, "Soft Corals," in *Coral Reefs: A Guide to the Common Invertebrates and Fishes of Bermuda, the Bahamas, Southern Florida, the West Indies, and the Caribbean Coast of Central and South America,* ed. Eugene H. Kaplan (Norwalk, CT: The Easton Press, 1982), 88.

9. Kaplan, *Coral Reefs,* 10.

10. Kinzie, "Soft Corals," 89.

11. Kinzie, "Soft Corals," 89.

12. Muzik, "Octocorallis," 305.

13. Muzik, "Octocorallis," 307.

14. Muzik, "Octocorallis," 305.

15. Karen Koltes, in-person interview, September 11, 2016.

16. Kinzie, "Soft Corals," 91–92.

17. Kinzie, "Soft Corals," 91.

18. Muzik, "Octocorallis," 306.

19. Kinzie, "Soft Corals," 91.

20. Kinzie, "Soft Corals," 92.

21. Greenberg, *Living Reef,* 20.

22. Burgess, *Corals,* 13.

23. Birkeland, "Introduction," 3.

24. Knowlton, *Citizens,* 81.

25. Burgess, *Corals,* 41–42.

26. Kaplan, *Coral Reefs,* 70.

27. Greenberg, *Living Reef,* 32.

28. Kaplan, *Coral Reefs,* 71.

29. Kaplan, *Coral Reefs,* 71–72.

30. Kaplan, *Coral Reefs,* 71.

31. Kaplan, *Coral Reefs,* 71.

32. Greenberg, *Living Reef,* 32.

33. University of California at Santa Barbara, UCSB Science Line. Last modified June 6, 2001, accessed January 24, 2019. https://www.scienceline.ucsb.edu.

34. Kaplan, *Coral Reefs,* 72.

35. Castro and Huber, *Marine Biology,* 118.

36. Castro and Huber, *Marine Biology,* 118–119.

37. Kaplan, *Coral Reefs,* 121.

38. Charles K. Biernbaum, "Invertebrate Experiments and Research Projects," 2. Accessed January 20, 2019. https://gricemarinelab.cofc.edu/education/teaching-resources/invertebrate-experiments.pdf.

39. Kaplan, *Coral Reefs,* 121.

40. Greenberg, *Living Reef,* 34.

41. Kaplan, *Coral Reefs,* 124.

42. Kaplan, *Coral Reefs,* 122–123.

43. Kaplan, *Coral Reefs,* 121.

44. Knowlton, *Citizens,* 118.

45. Kaplan, *Coral Reefs,* 125–126.

46. Burgess, *Corals,* 43.

47. Kaplan, *Coral Reefs,* 124.

48. Knowlton, *Citizens,* 36.

49. Knowlton, *Citizens,* 162.

50. Kaplan, *Coral Reefs,* 126

51. Knowlton, *Citizens,* 177.

52. Kaplan, *Coral Reefs,* 126.

53. Knowlton, *Citizens,* 53.

54. Sponge Docks. "Tarpon Springs Culture and History." Accessed January 21, 2019. https://spongedocks.net/tarpon-springs-history.htm.

55. John G. Macintyre, Klaus Rützler, James N. Norris, and Kristian Fauchald, "A Submarine Cave Near Columbus Cay, Belize: A Bizarre Cryptic Habitat," in Smithsonian Contributions to the Marine Sciences, no. 12, *The Atlantic Barrier Reef Ecosystem at Carrie Bow Cay, Belize, I: Structure and Communities,* eds. Klaus Rützler and Ian G. Macintyre (Washington, DC: Smithsonian Institution Press, 1982), 127.

56. Macintyre et al., "Submarine," 138.

t>3

57. Macintyre et al., "Submarine," 134.

58. Macintyre et al., "Submarine," 137.

59. Macintyre et al., "Submarine," 127.

60. Macintyre et al., "Submarine," 134.

61. Macintyre et al., "Submarine," 139.

62. Anja Schulze, phone interview, January 20, 2016.

63. Burgess, *Corals*, 43.

64. Greenberg, *Living Reef,* 40.

65. Schulze, phone interview, January 20, 2016.

66. Schulze, phone interview, January 20, 2016.

67. Schulze, phone interview, January 20, 2016.

68. Mary W. Rice and Ian G. Macintyre, "Distribution of Sipuncula in the Coral Reef Community, Carrie Bow Cay, Belize," in Smithsonian Contributions to the Marine Sciences, no. 12, *The Atlantic Barrier Reef Ecosystem at Carrie Bow Cay, Belize, I: Structure and Communities,* eds. Klaus Rützler and Ian G. Macintyre (Washington, DC: Smithsonian Institution Press, 1982), 311.

69. Rice and Macintyre, "Distribution," 318.

70. Paul G. Johnson and Barry A. Vittor, "Segmented Worms," in *Coral Reefs: A Guide to the Common Invertebrates and Fishes of Bermuda, the Bahamas, Southern Florida, the West Indies, and the Caribbean Coast of Central and South America,* ed. Eugene H. Kaplan (Norwalk, CT: The Easton Press, 1982), 134.

71. Rice and Macintyre, "Distribution," 318.

72. Johnson and Vittor, "Segmented," 136.

73. Kaplan, *Coral Reefs.* 36.

74. Schulze, phone interview, January 20, 2016.

75. Schulze, phone interview, January 20, 2016.

76. Schulze, phone interview, January 20, 2016.

Chapter 7. The Underwater Kaleidoscope: From Tiny Blennies to Giant Eagle Rays

1. Ron Eytan, in-person interview, January 6, 2016.

2. Belize.com, "Turneffe Atoll." Accessed January 7, 2019., https://belize.com/turneffe-atoll.

3. Kaplan, *Coral Reefs,* 230–231.

4. J. T. Williams and M. T. Craig, *"Ophioblennius atlanticus,"* *The IUCN Red List of Threatened Species 2014.* Accessed June 17, 2020. https://doi.org/10-2305/IUCN.UK.2014.

5. Kaplan, *Coral Reefs,* 230–231.

6. Ron Eytan, in-person interview, January 6, 2016.

7. David A. Greenfield and Teresa A. Greenfield, "Habitat and Resource Partitioning between Two Species of *Acanthemblemaria* (Pisces: Chaenopsidae), with Comments on the Chaos Hypothesis," in Smithsonian Contributions to the Marine Sciences, no. 12, *The Atlantic Barrier Reef Ecosystem at Carrie Bow Cay, Belize, I: Structure*

and *Communities*, eds. Klaus Rützler and Ian G. Macintyre (Washington, DC: Smithsonian Institution Press, 1982), 499.

8. Smithsonian Tropical Research Institute, "Species: *Acanthemblemaria aspera*, Roughhead Blenny." Posted 2015, accessed June 18, 2020. https://www.biogeodb.stri.edu/caribbean/en/thefishes/species/3990.

9. Eytan Lab webpage. Accessed January 27, 2019. https://www.roneytanlab.com.

10. Eytan Lab webpage.

11. Carole Baldwin, C. Castillo, L. Weight, and B. Victor, "Seven New Species within Western Atlantic *Starksia atlantica, S. lepicoela*, and *S. sluteri* (Teleostei, Labrisomidae), with Comments on Congruence of DNA Barcodes and Species," ZooKeys 79, 21–72. Posted 2011, accessed June 17, 2020. https://www.doi:10.3897/zookeys.79.1045.

12. Phil Hastings, phone interview, January 11, 2019.

13. Eytan, phone interview, February 25, 2019.

14. Eytan, in-person interview, January 6, 2016.

15. Eytan, phone interview, February 25, 2019.

16. Eytan, in-person interview, January 6, 2016.

17. Eytan Lab webpage. Accessed January 27, 2019. https://www.roneytanlab.com.

18. Eytan, phone interview, February 25, 2019.

19. Eytan, phone interview, February 25, 2019.

20. Eytan, in-person interview, January 6, 2016.

21. Hastings, phone interview, January 11, 2019.

22. Eytan, in-person interview, January 6, 2016.

23. Eytan, in-person interview, January 6, 2016.

24. Hastings, phone interview, January 11, 2019.

25. Ned and Anna DeLoach, "Leaping for Love," *Alert Diver* 34, no. 4, Fall 2018, 34–35.

26. Hastings, phone interview, January 11, 2019.

27. DeLoach, "Leaping," 216.

28. Kaplan, *Coral Reefs*, 207.

29. William Eschmeyer, *The Catalog of Fishes*, maintained online by the California Academy of Sciences. Accessed January 29, 2019. http://researcharchive.calacademy.org/research/ichthyology.

30. Ron Frank, William N. Eschmeyer, and Ray Van der Laan, eds., *Catalog of Fishes: Genera, Species, References*, 2018. Last modified January 2, 2019, accessed January 29, 2019. http://researcharchive.calacademy.org/research/ichthyology/catalog/fishcatmain.asp.

31. Connor J. Burgin, Jocelyn P. Colella, Philip L. Kahn, Nathan S. Upham, "How Many Species of Mammals Are There?" *Journal of Mammalogy 99*, no. 1, February 1, 2018, 1.

32. Smithsonian Institution, "Numbers of Insects (Species and Individuals)." Accessed June 17, 2020. https://www.si.edu/spotlight/buginfo/bugno.

33. David R. Bellwood and Peter C. Wainwright, "The History and Biogeography

of Fishes on Coral Reefs," in *Coral Reef Fishes: Dynamics and Diversity in a Complex Ecosystem*, ed. Peter F. Sale. (San Diego: Academic Press, 2002), 31.

34. Bellwood and Wainwright, "The History," 31.

35. Burgess, *Corals*, 46.

36. Coral Reef Alliance, "Coral Reef Ecology." Posted 2020, accessed June 18, 2020. https://www.coral.org/coral-reefs-101/coral-reef-ecology/.

37. Lourdes Vásquez Yeomans and Martha Elena Valdez Moreno. "Códigos de Barras de la Vida en Huevos y Larvas de Peces Costeros y Oceánicos de la Parte Norte del Sistema Arrecifal Mesoamericano (Caribe Mexicano)," 2009/12/15–2012/10/11. http://ecosur.mx/ecoconsulta/busqueda/detailes.

38. Florida Museum of Natural History, "Discover Fishes: Shark Biology." Accessed July 13, 2019. https://www.floridamuseum.ufl.edu/discover-fish/sharks/shark-biology/.

39. Peter Castro and Michael E. Huber, *Marine Biology: Eleventh Edition.* (New York: McGraw Hill Education, 2019), 167–168.

40. Knowlton, *Citizens*, 49.

41. Castro and Huber, *Marine Biology*, 161–162.

42. Castro and Huber, *Marine Biology*, 165–166.

43. Castro and Huber, *Marine Biology*, 163–164.

44. Peter C. Wainwright and David R. Bellwood, "Ecomorphology of Feeding in Coral Reef Fishes," in *Coral Reef Fishes: Dynamics and Diversity in a Complex Ecosystem*, ed. Peter F. Sale (San Diego: Academic Press, 2002). 33.

45. Wainwright and Bellwood, "Ecomorphology," 37–38.

46. Kaplan, *Coral Reefs*, 209.

47. Kaplan, *Coral Reefs*, 208.

48. Kaplan, *Coral Reefs*, 208.

49. Hastings, phone interview, January 11, 2019.

50. Kaplan, *Coral Reefs*, 208.

51. Burgess, *Corals*, 25.

52. Greenberg, *Living Reef*, 52.

53. Kaplan, *Coral Reefs*, 211.

54. Knowlton, *Citizens*, 100.

55. Lamar University, "Marine Critters." Accessed February 1, 2019, https://www.lamar.edu/marine-critters.

56. Florida Museum, "Discover Fishes: *Gymnothorax funebris*: Green Moray." Accessed June 18, 2020. https://www.floridamuseum.ufl.edu/species-profiles/gynmothorax-funebris/.

57. Lamar University, "Marine."

58. Castro and Huber, *Marine Biology*, 163.

59. Castro and Huber, *Marine Biology*, 340.

60. Kaplan, *Coral Reefs*, 260.

61. Knowlton, *Citizens*, 58.

62. *National Geographic*, "Pufferfish." Last modified 2015, accessed June 18, 2020. https://www.nationalgeographic.com/animals/fish/puggerfish/.

63. Jorge Lago, Laura P. Rodriguez, Lucia Blanco, Juan Manuel Vieites, and Ana G. Cabado, "Tetrodotoxin, an Extremely Potent Marine Neurotoxin: Distribution, Toxicity, Origin and Therapeutic Uses," *Marine Drugs* 13, no. 10, 2015, 6384–6406.

64. Lago et al., "Tetrodotoxin," 6384–6406.

65. Knowlton, *Citizens*, 167.

66. Kaplan, *Coral Reefs*, 261.

67. Kaplan, *Coral Reefs*, 261.

68. Knowlton, *Citizens*, 167.

69. Fugu Fukuji website. Accessed January 30, 2019. https://www.fukuji.jp.

70. Knowlton, *Citizens*, 167.

71. Florida Museum, "Discover Fishes: Spotted Scorpionfish." Accessed June 19, 2020. https://www.floridamuseum.ufl.edu/discover-fish/species-profiles/scorpaena-plumieri/.

72. Greenberg, *Living Reef*, 62.

73. Sarah Egner, "Parrotfish: Grazers of the Reef," *Alert Diver 32*, no. 3, Summer 2016, 38.

74. Knowlton, *Citizens*, 35.

75. Kaplan, *Coral Reefs*, 212.

76. Kaplan, *Coral Reefs*, 209.

77. Knowlton, *Citizens*, 45.

78. Georgia Aquarium, "Animal Guide: Porkfish." Accessed January 30, 2019, https://www.georgiaaquarium.org/animal/porkfish.

79. Knowlton, *Citizens*, 29.

80. Kaplan, *Coral Reefs*, 256–257.

81. Saint Louis Zoo, "Southern Stingray Facts." Accessed July 14, 2019. https://www.stlzoo.org/visit/thingstoseeanddo/stingraysatcaribbeancove/southernstingrayfacts.

82. Florida Museum, "Discover Fishes: *Aetobatus narinari*: Spotted Eagle Ray." Accessed June 19, 2020. https://www.floridamuseum.ufl.edu/species-profiles/aetobatus-narinari/.

83. Associated Press, "Ray Leaps from Water, Killing a Boater in the Florida Keys." Last modified March 21, 2008, accessed July 14, 2019. https://www.nytimes.com/2008/03/21/us/21sting.html.

84. Robert Lewis, "Manta Ray," *Encyclopaedia Britannica*. Accessed July 14, 2019. https://www.britannica.com/animal/manta-ray/.

85. Shark Research Institute, "Catching Some Rays in Mexico." Accessed July 14, 2019. https://www.sharks.org/shark-research-institute-blogs/blogs/science-blog/catching-rays-mexico.

86. Greenberg, *Living Reef*, 122–123.

87. Knowlton, *Citizens*, 45.

88. Greenberg, *Living Reef*, 58.

89. Greenberg, *Living Reef*, 54.

90. Knowlton, *Citizens*, 45.

91. Knowlton, *Citizens*, 45.

92. Knowlton, *Citizens*, 73.

93. Knowlton, *Citizens*, 46.

94. Knowlton, *Citizens*, 64.

95. Kaplan, *Coral Reefs*, 209.

96. Kaplan, *Coral Reefs*, 207.

97. Elizabeth Preston, "Happy Father's Day to All the Fish Dads Underwater," *New York Times*. Posted June 21, 2020, accessed June 21, 2020. https://www.nytimes.com/article/fathers-day-fish.html.

98. Knowlton, *Citizens*, 146.

99. Greenberg, *Living Reef*, 88.

100. Knowlton, *Citizens*, 144.

101. Greenberg, *Living Reef*, 102.

102. Greenberg, *Living Reef*, 102

103. Greenberg, *Living Reef*, 119.

104. Eytan, in-person interview, January 6, 2016.

105. Brandon Cole, "Cozumel Revisited," *Alert Diver 34*, no. 3, Summer 2018, 79.

106. Diego Barneche and Dustin Marshall, "Fishy Limits," *Alert Diver 35*, no. 1, Winter 2019, 110.

107. Barneche and Marshall, "Fishy," 112.

108. Loren McLenachan, "Documenting Loss of Large Trophy Fish from the Florida Keys with Historical Photographs," *Conservation Biology 23*, no. 3, 2009, 636–643.

109. Therese Bowman Rath, phone interview, October 11, 2016.

110. Bowman Rath, phone interview, October 11, 2016.

111. Tschirky, in-person interview, September 8, 2016.

Chapter 8. Big, Bold, and Vanishing: Groupers and Parrotfishes

1. Florida Fish and Wildlife Conservation Commission, "Goliath Grouper." Accessed June 20, 2020. https://myfwc.com/fishing/saltwater/recreational/goliath/.

2. Florida Museum, "Nassau Grouper." Accessed June 20, 2020. https://www.floridamuseum.ufl.edu/discover-fish/species-profiles/epinephelus-striatus/.

3. Knowlton, *Citizens*, 118.

4. Marine Conservation Institute, "Glover's Reef Marine Reserve," *Atlas of Marine Protection*. Posted 2020, accessed June 20, 2020. https://mpatlas.org.

5. National Oceanic and Atmospheric Administration, "Species Directory: Black Grouper." Accessed January 31, 2019. https://www.fisheries.noaa.gov/species/black-grouper. Ned and Anna DeLoach, "How to Save a Fish," *Alert Diver 34*, no. 3, Summer 2018, 32.

6. Roberto Pott, phone interview, October 12, 2016.

7. Pott, phone interview, October 12, 2016.

8. Knowlton, *Citizens,* 139.

9. DeLoach, "How to Save," 32.

10. Knowlton, *Citizens,* 139.

11. DeLoach, "How to Save ," 32–33.

12. Knowlton, *Citizens,* 139.

13. National Oceanic and Atmospheric Administration, "Species Directory: Nassau Grouper." Accessed January 31, 2019. http://www.fisheries.noaa.gov/species/nassaugrouper.

14. DeLoach, "How to Save," 32.15. *National Geographic,* "Cubera Snapper." Accessed June 6, 2020. https://www.nationalgeographic.com/animals/fish/cubera-snapper/.

16. Florida Fish and Wildlife Conservation Commission, "Species Profiles." Accessed February 1, 2019. https://myfwc.com/wildlifehabitats/profiles.

17. Rowe, in-person interview, January 7, 2016.

18. Greenberg, *Living Reef,* 123.

19. Sarah Egner, "Parrotfish: Grazers of the Reef," *Alert Diver 32,* no. 3, Summer 2016, 38.

20. J. R. Caldwell and N. R. Liley, "Hormonal Control of Sex and Color in the Stoplight Parrotfish, *Sparisoma Viride,*" *General and Comparative Endocrinology 81,* no. 1, January 1991, 7–20.

21. Laura B. Carano, Bridgette K. Gunn, Megan C. Kelly, and Deron E. Burkepile, "Predation Risk, Resource Quality, and Reef Structural Complexity Shape Territoriality in a Coral Reef Herbivore," *PLoS One 10,* no. 2, 2015. Posted February 25, 2015, accessed June 20, 2020. https://www.ncbi.nlm.nih.gov/pmc/articles/PMC4340949/.

22. Egner, "Parrotfish," 38.

23. Mireille L. Harmelin-Vivien, "Energetics and Fish Diversity on Coral Reefs," in *Coral Reef Fishes: Dynamics and Diversity in a Complex Ecosystem,* ed. Peter F. Sale (San Diego: Academic Press, 2002), 268.

24. Egner, "Parrotfish," 40.

25. Egner, "Parrotfish," 39.

26. Melanie McField, phone interview, October 11, 2016.

27. Rueda Flores, phone interview, November 6, 2016.

28. McField et al., *Report Card 2020,* 17.

29. Ana Giró Petersen, phone interview, November 6, 2016.

30. Giró Petersen, phone interview, November 6, 2016.

31. McField et al., *Mesoamerican Reef,* 6.

32. Macrotrends, "Mexico Population 1950–2021." https://www. .macrotrends.net/countries/MEX/mexico/population. Posted January 2021, accessed February 17, 2021. https://www.macrotrends.net/countries/MEX/mexico/population.

33. Countrymeters. Accessed February 6, 2019. https://www.countrymeters.info.

34. Rueda Flores, phone interview, November 3, 2016.

Chapter 9. Where Have All the Sea Urchins Gone?

1. Names.org. Accessed February 7, 2019. https://www.names.org.

2. Reefguide.org. Accessed January 8, 2019. https://www.reefguide.org/carib/in-dex22.html.

3. Castro and Huber, *Marine Biology*, 145.

4. Knowlton, *Citizens*, 42.

5. Kaplan, *Coral Reefs*, 188.

6. Castro and Huber, *Marine Biology*, 145.

7. Kaplan, *Coral Reefs*, 188.

8. Kaplan, *Coral Reefs*, 187.

9. Reefguide.org. Accessed February 8, 2019. http://www.reefguide.org/carib/in-des22.html.

10. Philip James and Sten Siikavuopio, *A Guide to the Sea Urchin Reproductive Cycle and Staging Sea Urchin Gonad Samples* (Tromsø, Norway: Nofima AS), 5. Accessed June 21, 2020. https://www.nofima.no/filearchive/guide-to-sea-urchins_lowres.pdf.

11. Kellie C. Pelikan, "The Effects of Petroleum Pollutants on Sea Urchins Reproduction and Development," master's thesis. Nova Southeastern University, submitted December 7, 2015. Accessed February 9, 2019. https://nsuworks.nova.edu/occ_stuetd/401.

12. Kaplan, *Coral Reefs*, 186.

13. Kaplan, *Coral Reefs*, 188.

14. James Engman, "Mass Mortality in *Diadema antillarum* (*Echinodermata: Echinoidea*): A Large-Scale Natural Experiment in Herbivore Removal." Last modified October 29, 2002, accessed February 9, 2019. https://www.jrscience.wcp.muchio.edu.

15. Harilaos Lessios, phone interview, February 11, 2019.

16. Lessios, phone interview, February 11, 2019.

17. Lessios, phone interview, February 11, 2019.

18. Engman, "Mass Mortality."

19. Lessios, phone interview, February 11, 2019.

20. Harilaos A. Lessios, John D. Cubit, and D. Ross Robertson, "Spread of *Diadema* Mass Mortality through the Caribbean," *Science 226*, November 1984, 335.

21. Harilaos Lessios, "The Great *Diadema antillarum* Die-Off: 30 Years Later, *Annual Review of Marine Sciences 8*, January 2016, 268. First published online as a Review in Advance on June 5, 2015. https://doi.org/10.1146/annurev-marine-122414-033857.

22. Lessios et al., "Spread," 335.

23. Bayard Webster, "Sea Urchin Deaths Puzzle Scientists," *New York Times*. Last modified October 12, 1984, accessed February 9, 2019. https://www.nytimes.com/1984/10/12/us/sea-urchin-deaths-puzzle-scientists.html.

24. Nancy Knowlton, phone interview, January 4, 2019.

25. Knowlton, phone interview, February 4, 2019.

26. Knowlton, phone interview, January 4, 2019.

27. Lessios, phone interview, February 11, 2019.

28. Engman, "Mass Mortality."

29. Harilaos Lessios, "The Great *Diadema antillarum* Die-Off: 30 Years Later, *Annual Review of Marine Sciences 2016 8*, June 5, 2015, 268.

30. Lessios, "Great," 268.

31. Lessios, "Great," 278.

32. Engman, "Mass Mortality."

33. Nancy Knowlton, "Sea Urchin Recovery from Mass Mortality: New Hope for Caribbean Coral Reefs?" *Proceedings of the National Academy of Sciences of the United States,* April 24, 2001, 4824.

34. Robin Meadows, "Overfishing Implicated in Sea Urchin Epidemics," *Conservation Magazine,* University of Washington. Last modified July 29, 2008, accessed February 11, 2019. http://www.conservationmagazine.org/2008/07/overfishing-implicated-in-sea-urchin-epidenics.

35. Knowlton, "Sea Urchin," 4822–4824.

36. James N. Norris and William Fenical, "Chemical Defense in Tropical Marine Algae," in Smithsonian Contributions to the Marine Sciences, no. 12, *The Atlantic Barrier Reef Ecosystem at Carrie Bow Cay, Belize, I: Structure and Communities*, eds. Klaus Rützler and Ian G. Macintyre (Washington, DC: Smithsonian Institution Press, 1982), 427.

37. Lessios, "Great," 270.

38. Lessios, "Great," 272.

39. Lessios, "Great," 272.

40. Lessios, "Great," 273.

41. Knowlton, "Sea Urchin," 4823.

42. Lessios, "Great," 275.

43. Knowlton, "Sea Urchin," 4823.

44. Knowlton, "Sea Urchin," 4824.

45. Knowlton, "Sea Urchin," 4824.

46. Lessios, phone interview, February 11, 2019.

47. Lessios, "Great," 275–276.

48. Lessios, "Great," 276.

49. Lessios, "Great," 276.

50. Lessios, "Great," 276.

51. Lessios, "Great," 273.

52. Knowlton, "Sea Urchin," 4823.

53. McField et al., *Mesoamerican Reef,* 14.

54. Peter Edmunds, phone interview, February 22, 2019.

55. Peter J. Edmunds and Robert C. Carpenter, "Recovery of *Diadema antillarum* Reduces Macroalgae Cover and Increases Abundance of Juvenile Corals in a Caribbean Reef," *Proceedings of the National Academy of Sciences 98*, no. 9, May 2001, 5067–5071.

56. Edmunds, phone interview, February 22, 2019.

57. Edmunds, phone interview, February 22, 2019.

58. Edmunds, phone interview, February 22, 2019.

59. Engman, "Mass Mortality."

60. Lessios, "Great," 277.

61. Engman, "Mass Mortality."

62. Engman, "Mass Mortality."

63. Webster, "Sea Urchin."

64. Lessios, phone interview, February 11, 2019.

Chapter 10. Starring Roles: Starfishes and Their Kin

1. Knowlton, *Citizens,* 118.

2. Kaplan, *Coral Reefs,* 169.

3. Castro and Huber, *Marine Biology,* 145.

4. Kaplan, *Coral Reefs,* 173–176.

5. Stephen Gittings, written correspondence, October 7, 2019.

6. Desmond Brown, "Caribbean Fears Loss of 'Keystone Species' to Climate Change," Inter Press Service. Accessed March 22, 2019. http://www.ipsnews.net/2014 /04/caribbean-fears-loss-keystone-species-climate-change/.

7. Detecon Consulting, "Speaking of Starfish and Mussels: Role Distribution and Success Strategies in Business Ecosystems." Accessed March 24, 2019. https://www. detecon.com/en/knwledge/speaking-stafish-and-mussels.

8. Detecon, "Speaking."

9. Castro and Huber, *Marine Biology,* 144.

10. Kaplan, *Coral Reefs,* 169.

11. Kaplan, *Coral Reefs,* 170.

12. Kaplan, *Coral Reefs,* 171.

13. Kaplan, *Coral Reefs,* 172.

14. Knowlton, *Citizens,* 46.

15. Kaplan, *Coral Reefs,* 172.

16. Kaplan, *Coral Reefs,* 173.

17. Kaplan, *Coral Reefs,* 171.

18. Kaplan, *Coral Reefs,* 175–176.

19. Kaplan, *Coral Reefs,* 171.

20. Gordon T. Taylor and Richard D. Bray, "Brittle or Serpent Stars," in *Coral Reefs; A Guide to the Common Invertebrates and Fishes of Bermuda, the Bahamas, Southern Florida, the West Indies, and the Caribbean Coast of Central and South America,* ed. Eugene H. Kaplan (Norwalk, CT: The Easton Press, 1982), 177.

21. Knowlton, *Citizens,* 77.

22. Juan José Alvarado, "Echinoderm Diversity in the Caribbean Sea," *Marine Biodiversity 41,* 2011, 261–285. Published July 22, 2010, accessed June 6, 2020. https://www. springer.com/article/10.1007/s12526-010-0053-0.

23. Christopher M. Pomory, "Key to Common Shallow-Water Brittle Stars (*Echinodermata: Ophiuroidea*) of the Gulf of Mexico and Caribbean Sea," *Caribbean Journal*

of Science 10. Published January 2007, accessed June 6, 2020. https://www.research-gate.net/publication/228496999_K_to_the_common_shallow-water_brittle_stars_Echinodermata_Ophiuroidea_of_the_Gulf_of_Mexico_and_Caribbean_Sea.

24. Taylor and Bray, "Brittle," 177.

25. Taylor and Bray, "Brittle," 177.

26. Taylor and Bray, "Brittle," 177.

27. Taylor and Bray, "Brittle," 178.

28. Taylor and Bray, "Brittle," 178.

29. Taylor and Bray, "Brittle," 178.

30. Kaplan, *Coral Reefs,* 194.

31. Marty Snyderman and Clay Wiseman, *Guide to Marine Life: Caribbean, Bahamas, Florida* (New York: Aqua Quest Publications, 1996), 125.

32. Snyderman and Wiseman, *Guide,* 125.

33. Kaplan, *Coral Reefs,* 195.

34. Matt Simon, "Absurd Creature of the Week: This Fish Swims up a Sea Cucumber's Butt and Eats Its Gonads," *Wired.* Last modified February 21, 2014, accessed March 26, 2019. https://www.wired.com/2014/02/absurd-creature-of-the-week-pearlfish/.

35. Simon, "Absurd."

36. Kaplan, *Coral Reefs,* 196.

37. Richard T. Bray and Gordon T. Taylor, "Sea Lilies or Feather Stars," in *Coral Reefs; A Guide to the Common Invertebrates and Fishes of Bermuda, the Bahamas, Southern Florida, the West Indies, and the Caribbean Coast of Central and South America,* ed. Eugene H. Kaplan, (Norwalk, CT: The Easton Press, 1982), 202.

38. Castro and Huber, *Marine Biology,* 147.

39. Bray and Taylor, "Sea Lilies," 203.

40. Bray and Taylor, "Sea Lilies," 203.

41. Bray and Taylor, "Sea Lilies," 204.

Chapter 11. Hiding Out: Creatures at Home inside the Reef

1. Malcolm Telford, "Shrimps, Lobsters, and Crabs," in *Coral Reefs: A Guide to the Common Invertebrates and Fishes of Bermuda, the Bahamas, Southern Florida, the West Indies, and the Caribbean Coast of Central and South America,* ed. Eugene H. Kaplan, (Norwalk, CT: The Easton Press, 1982), 157.

2. National Oceanic and Atmospheric Administration, "Caribbean Spiny Lobster." Accessed June 27, 2020. https://www.fisheries.gov/species/caribbean-spiny-lobster.

3. Castro and Huber, *Marine Biology,* 140.

4. University of Leicester, "Lobster Telescope Has an Eye for X-Rays," *Science Daily,* April 5, 2006. https://www.sciencedaily.com/releases/2006/04/060404194138.htm.

5. Telford, "Shrimps, Lobsters," 157.

6. Erica Ross and Donald C. Behringer, "Changes in Temperature, pH, and

Salinity Affect the Sheltering Responses of Caribbean Spiny Lobsters to Chemosensory Cues," *Scientific Reports 9,* article 4375, March 13, 2019.

7. Mihika T. Kozma, Manfred Schmidt, Hanh Ngo-Vu, Shea D. Sparks, Adriane Senatore, and Charles D. Derby, "Chemoreceptor Proteins in the Caribbean Spiny Lobster, *Panulirus argus:* Expression of Ionotropic Receptors, Gustatory Receptors, and TRP Channels in Two Chemosensory Organs and Brain," *PLoS One 13,* no. 9, 2018. Posted online September 21, 2018, accessed June 27, 2020. https://www.ncbi.nlm.nih.gov/pmc/articles/PMC6150509/.

8. Dana M. Hawley and Julia C. Buck, "Animals Use Social Distancing to Avoid Disease," originally published as "Animals Apart" in *Scientific American 323,* no. 2, 36–41, August 2020. http://doi.org/10.1038/scientificamerican0820-36.

9. Telford, "Shrimps, Lobsters," 157.

10. P. E. Bouwma and W. F. Herrnkind, "Spiny Lobsters Combine Weaponry with Sound to 'Teach' Predators Not to Attack," 34th Annual Benthic Ecology Meeting, Williamsburg, VA, April 6–10, 2005.

11. Kim Ley-Cooper, *Sustainability of Lobster* Panulirus argus *Fisheries in Marine Protected Areas in South-eastern Mexico,* doctoral thesis. Curtin University, Perth, Australia, submitted December 2015, 83.

12. National Oceanic and Atmospheric Administration, "Caribbean Spiny Lobster."

13. National Oceanic and Atmospheric Administration, "Caribbean Spiny Lobster."

14. Litzy Ayra and Raúl Cruz, "Caribbean Lobster Reproduction: A Review," *Revista Investigaciones Marinas 3,* no. 2, 2010, 116.

15. National Oceanic and Atmospheric Administration, "Caribbean Spiny Lobster."

16. National Oceanic and Atmospheric Administration, "Caribbean Spiny Lobster."

17. Nadine Seudeal, *"Panulirus argus:* Caribbean Spiny Lobster," Animal Diversity Web, University of Michigan Museum of Zoology. Posted 2020, accessed June 27, 2020. https://animaldiversity.org/accounts/Panulirus_argus/.

18. Jason S. Goldstein, H. Matsuda, T. Takenouchi, and Mark J. Butler IV, "The Complete Development of Larval Caribbean Spiny Lobster *Panulirus argus* (Latreille, 1804) in Culture," *Journal of Crustacean Biology 28,* no. 2, 2008, 307.

19. National Oceanic and Atmospheric Administration, "Caribbean Spiny Lobster."

20. Jason S. Goldstein and Mark J. Butler IV, "Behavioral Enhancement of Onshore Transport by Postlarval Caribbean Spiny Lobster *(Panulirus argus),*" *Limnology and Oceanography 54,* no. 3, 2009, 1669.

21. Goldstein and Butler, "Behavioral Enhancement," 1669.

22. National Oceanic and Atmospheric Administration, "Caribbean Spiny Lobster."

23. Nadine Seudeal, *"Panulirus argus."*

24. Snyderman and Wiseman, *Guide,* 108.

25. National Oceanic and Atmospheric Administration, "Caribbean Spiny Lobster."

26. Mark J. Butler, Donald C. Behringer, and Jeffrey D. Shields, "Transmission of Panulirus Argus Virus 1 (PaV1) and Its Effect on the Survival of Juvenile Caribbean Spiny Lobster," *Diseases of Aquatic Organisms 79*, no. 3, May 8, 2008, 173.

27. National Oceanic and Atmospheric Administration, "Caribbean Spiny Lobster."

28. Knowlton, *Citizens,* 127.

29. Knowlton, *Citizens,* 127.

30. Donald C. Behringer, Mark J. Butler IV, and Grant D. Stentiford, "Disease Effects on Lobster Fisheries, Ecology, and Culture: Overview of DAO Special 6," *Diseases of Aquatic Organisms* 100, 2012, 90.

31. Knowlton, *Citizens,* 25.

32. Marine Education Society of Australasia, "Crustaceans." Last modified June 25, 2019, accessed June 25, 2019. https://www.mesa.edu.au/crustaceans/.

33. Snyderman and Wiseman, *Guide,* 110.

34. Snyderman and Wiseman, *Guide,* 110.

35. McField et al., *Report Card 2020,* 15.

36. Encyclopaedia Britannica online, "Coral Gall Crab." Accessed July 3, 2019. https://www.britannica.com/animals/coral-gall-crab.

37. R. Johnsson, E. Neves, G.M.O. Franco, and F. L. Silveira, "The Association of Two Gall Crabs (*Brachyura: Cryptochiridae*) with the Reef-Building Coral *Siderastrea stellata* Verrill, 1868," *Hydrobiologia 559*, no. 1, 2006, 383.

38. Burgess, *Corals,* 44.

39. Johnsson et al., "The Association," 379.

40. Telford, "Shrimps, Lobsters," 151.

41. Telford, "Shrimps, Lobsters," 165.

42. Telford, "Shrimps, Lobsters," 151.

43. Knowlton, *Citizens,* 91.

44. Telford, "Shrimps, Lobsters," 158–159.

45. Telford, "Shrimps, Lobsters," 158.

46. Melissa Block, "*Stenorhynchus seticornis:* Yellowline Arrow Crab," Animal Diversity Web. Accessed July 3, 2019. https://www.animaldiversity.org/accounts/Stenorhynchus_seticornis.

47. Telford, "Shrimps, Lobsters," 151–153.

48. Jill Kokemuller, "What Are Shrimps' Prey?" Last modified August 6, 2018, accessed July 4, 2019. https://sciencing.com/shrimps-prey-8309570.html.

49. Telford, "Shrimps, Lobsters," 155.

50. Knowlton, *Citizens,* 128.

51. Knowlton, *Citizens,* 110.

52. Knowlton, *Citizens,* 115.

53. Ned and Anna DeLoach, "Pest Control: Part Two: The Manicure," *Alert Diver 35,* no. 2, Spring 2019, 32–33.

54. DeLoach, "Pest," 32–33.

55. Knowlton, *Citizens,* 115.

56. Telford, "Shrimps, Lobsters," 153.

57. Knowlton, *Citizens,* 68.

58. Knowlton, *Citizens,* 96.

59. Leah Crane, "Mantis Shrimp Punch with the Force of a Bullet—and Now We Know How," *New Scientist.* Posted October 18, 2018, accessed June 7, 2020. https://www.newscientist.com/article/2188880-mantis-shrimps-punch-with-the-force-of-a-bullet-andnow-we-know-how.

60. Telford, "Shrimps, Lobsters," 153.

61. Knowlton, *Citizens,* 42.

62. *Brevard Times,* "Giant," November 25, 2015.

Chapter 12. Armed for Survival: Octopuses and Squids

1. Gilbert L. Voss and Clyde F. E. Roper, "Cephalopod Class of Mollusks," Encyclopaedia Britannica online. Last modified March 3, 20117, accessed September 1, 2017. https://www.britannica.com/animal/cephalopod.

2. Voss and Roper, "Cephalopod."

3. Voss and Roper, "Cephalopod."

4. Voss and Roper, "Cephalopod."

5. R. Toonen, "Aquarium Invertebrates: Housing an Octopus," *Advanced Aquarist* II. Accessed July 2013. https://www.advancedaquarist.com.

6. Bernd U. Budelmann, "The Cephalopod Nervous System: What Evolution Has Made of the Molluscan Design," in *The Nervous System of Invertebrates: An Evolutionary and Comparative Approach,* eds. O. Breidbach and W. Kutsch (Basel, Switzerland: Birkhauser Verlag, 1995), 115.

7. Budelmann, "Cephalopod Nervous," 115.

8. Bernd U. Budelmann, phone interview, February 16, 2016.

9. Bernd U. Budelmann, "Active Marine Predators: The Sensory World of Cephalopods," *Marine and Freshwater Behaviour and Physiology* 27, nos. 2–3, 1996, 59.

10. Budelmann, "Cephalopod Nervous," 132.

11. Budelmann, "Active," 60.

12. Voss and Roper, "Cephalopod."

13. Voss and Roper, "Cephalopod."

14. Voss and Roper, "Cephalopod."

15. Budelmann, "Cephalopod Nervous," 132.

16. Budelmann, "Active," 64.

17. Budelmann, "Active," 64.

18. Budelmann, "Active," 64.

19. Budelmann, "Active," 64.

20. Suzana Herculano-Hourzel, "The Human Brain in Numbers: A Linearly Scaled-up Primate Brain," *Frontiers in Human Neuroscience* 3, no. 31. Posted November

9, 2009, accessed June 29, 2020. https://www.frontiersin.org/articles/10.3389/neuro.09.031.2009/full.

21. Budelmann, "Cephalopod Nervous," 116.

22. Budelmann, "Cephalopod Nervous," 116.

23. Lena van Giersen, Peter B. Kilian, Corey A. H. Allard, and Nicolas W. Bellono, "Molecular Basis of Chemotactile Sensation in Octopus," *Cell 183,* no. 3, October 29, 2020, 594–604.

24. Budelmann, "Cephalopod Nervous," 118–19.

25. Maggy Wassilieff and Steve O'Shea, "Octopus and Squid—Feeding and Predation," *Te Ara—the Encyclopedia of New Zealand.* Last modified June 12, 2006, accessed September 1, 2017. http://www.TeAra.govt.nz/en/octopus-and-squid/page-3.

26. Voss and Roper, "Cephalopod."

27. Budelmann, phone interview, February 12, 2016.

28. Budelmann, phone interview, February 12, 2016.

29. Budelmann, phone interview, February 12, 2016,

30. Budelmann, phone interview, February 12, 2016.

31. National Geographic, "Common Octopus." Accessed June 8, 2020. https://www.nationalgeographic.com/animals/invertebrates/common-octopus/.

32. Phil Myers, "Caribbean Reef Octopus," *Animal Diversity,* University of Michigan Museum of Zoology. Posted 2001, accessed June 8, 2020. https://www.animaldiversity.org/accounts/Octopus_briareus/.

33. Budelmann, in-person interview, October 2, 2015.

34. Judith Martin, "A Ring around the Mood Market," *Washington Post,* November 24, 1975, B9.

35. Roland C. Anderson, Jennifer A. Martin, and James B. Wood, *Octopus: The Ocean's Intelligent Invertebrate* (Portland, OR: Timber Press, 2013), 147.

36. Budelmann, phone interview, February 16, 2016.

37. Budelmann, "Cephalopod Nervous," 131.

38. Budelmann, "Active Marine," 62.

39. Godfrey-Smith, Peter, *Other Minds: The Octopus, the Sea, and the Deep Origins of Consciousness* (New York: Farrar, Strauss and Giroux, 2016), 5.

40. Godfrey-Smith, *Other Minds,* 9.

41. Godfrey-Smith, *Other Minds,* 9.

42. Budelmann, phone interview, February 12, 2016.

43. Roger Hanlon, "Nature's Best and Fastest Camouflage," *Alert Diver 33,* no. 4, Fall 2017, 42–44.

44. Hanlon, "Nature's Best," 43.

45. Budelmann, "Cephalopod Nervous," 133.

46. Budelmann, in-person interview, October 2, 2015.

47. Sy Montgomery, *The Soul of an Octopus: A Surprising Exploration into the Wonder of Consciousness* (New York: Atria Books, 2015), 228.

48. Budelmann, phone interview, February 12, 2016.

49. Budelmann, phone interview, February 12, 2016.

Chapter 13. Aquatic Intellectuals: Dolphins

1. Eytan, in-person interview, January 6, 2016.

2. Berkeley Evolibrary, "The Evolution of Whales." Accessed July 7, 2019. https://evolution.berkeley.edu/evolibrary/article/evograms_03.

3. Joshua Foer, "It's Time for a Conversation: Breaking the Communication Barrier between Dolphins and Humans," *National Geographic,* May 2015, 37.

4. National Ocean Service, National Oceanic and Atmospheric Administration, "What's the Difference between Dolphins and Porpoises?" Accessed July 7, 2019. https://oceanservice/noaa.gov/facts/dolphin_porpoise.html.

5. Foer, "It's Time," 36.

6. Foer, "It's Time," 51.

7. Foer, "It's Time," 44.

8. Knowlton, *Citizens,* 53.

9. Knowlton, *Citizens,* 53.

10. Foer, "It's Time," 54.

11. U.S. Whales, "How Do Dolphins Give Birth?" Accessed July 7, 2019. https://us.whales.org/whales-dolphins/how-do-dolphins-give-birth/.

12. Sea World Parks and Entertainment. Accessed July 9, 2019. https://seaworld.org/animals/all-about/bottlenose-dolphins/longevity/.

13. Max Roser, "Child and Infant Mortality," Our World in Data. Accessed July 9, 2019. https://ourworldindata.org/child-mortality.

14. U.S. Whales, "How Do Dolphins Give Birth?"

15. Foer, "It's Time," 46–47.

16. San Diego Zoo, "Dolphin." Accessed July 7, 2019. https://animals.sandiegozoo.org/animals/dolphin.

17. Foer, "It's Time," 37.

18. Foer, "It's Time," 40–42.

19. Foer, "It's Time," 47–49.

20. Foer, "It's Time," 43–44.

21. Foer, "It's Time," 47.

22. Foer, "It's Time," 44.

23. Erica Tennenhouse, "These Fishermen Helping Dolphins Have Their Own Culture," *National Geographic.* Last modified April 9, 2019, accessed July 8, 2019. https://www.nationalgeographic.com/animals/2019/04/dolphins-fishermen-brazil-culture/.

24. Lina Zeldovich, "The Great Dolphin Dilemma," *Hakai Magazine,* February 5, 2019. Accessed July 7, 2019. https://wwwhakaimagazine.com/features/the-great-dolphin-dilemma.

25. Foer, "It's Time," 36.

26. Zeldovich, "Great Dolphin."

27. Foer, "It's Time," 36.

28. Joseph Castro, "How Do Dolphins Sleep?" *Live Science.* Last modified April

14, 2014, accessed July 8, 2019. https://www.livescience.com/how-do-dolphins-sleep/html.

29. U.S. Whales, "How Do Dolphins Sleep?" Accessed July 8, 2019. https://us.whales.org/whales-dolphins/how-do-dolphins-sleep/.

30. Foer, "It's Time," 43.

31. *Blue Planet II* (United Kingdom: BBC Natural History Unit, 2017).

32. Foer, "It's Time," 52.

33. Foer, "It's Time," 52.

34. Foer, "It's Time," 55.

35. Stan A. Kuczaj II and Deidre Yeater, "Dolphin Imitation: Who, What, When, and Why?" *Aquatic Mammals* 32, no. 4, 2006, 413.

36. Stan A. Kuczaj and Holli C. Eskelinen, "Why Do Dolphins Play?" *Animal Behavior and Cognition* 1, no. 2, 2014, 113.

37. Foer, "It's Time," 47 and 51.

38. *Washington Post*, "Inventor Studied Dolphin Communications." Last modified October 4, 2001, accessed July 10, 2019. https://www.washingtonpost.com/archive/local/2001/10/04/john-c-lilly/594dc878-43ac-.

39. Foer, "It's Time," 47.

40. *Washington Post*, "Inventor."

41. *Washington Post*, "Inventor."

42. *Washington Post*, "Inventor."

43. Foer, "It's Time," 35.

44. Dolphin Communication Project. Accessed July 10, 2019. https://www.dolphincommunicationproject.org/index.php/the-latest-buzz/field-reports/bahamas-3/bahamas-2000/item/94230-*stan-kuczaj-ph-d*.

45. Foer, "It's Time," 52.

46. Foer, "It's Time," 52.

47. Foer, "It's Time," 53.

48. Foer, "It's Time," 54.

49. Douglas Main, "Why Koko the Gorilla Mattered," *National Geographic*. Last modified June 21, 2018, accessed July 8, 2019. https://www.news.nationalgographic.com/2018/06/gorillas-koko-sign-language-culture-animals.

50. Main, "Why Koko."

51. Ulfur Arnason, Annette Gullberg, and Axel Janke, "Molecular Timing of Primate Divergences as Estimated by Two Nonprimate Calibration Points," *Journal of Molecular Evolution* 47, no. 6, December 1998, 718–727.

52. Foer, "It's Time," 45.

53. Diana Kwon, "What Makes Our Brains Special?" *Scientific American*, November 24, 2015. Accessed July 11, 2019. https://www.scientificamerican.com/article/what-makes-our-brains-special/.

54. Yuval Noah Harari, *Sapiens: A Brief History of Humankind* (New York: Harper Perennial, 2015), 28.

55. Foer, "It's Time," 43.

56. Foer, "It's Time," 43.

Chapter 14. Misunderstood "Bad Boys of the Reef": Sharks

1. Elaina Zachos, "Why Are We Afraid of Sharks? There's a Scientific Explanation," *National Geographic*. Last modified June 27, 2019, accessed July 12, 2019. https://news.nationalgeographic.com/2018/01/sharks-attack-fear-science-psychology-spd/.

2. Josh Davis, "Shark Evolution: A 450 Million Year Timeline," Natural History Museum. Last modified December 13, 2018, accessed July 16, 2019. https://www.nhm.ac.uk/discover/shark-evolution-a-450-million-year-timeline.html.

3. Zachos, "Why Are."

4. Sarah Zeilinski, "What Preys on Humans?" *Smithsonian Magazine*. Last modified July 22, 2011, accessed July 16, 2019. https://www.smithsonianmag.com/science-nature/what-preys-on-humans-34332952/.

5. Zachos, "Why Are."

6. Maya Wei-Haas, "One of the Biggest Great White Sharks Feasting on a Sperm Whale in Rare Video," *National Geographic*. Last modified July 19, 2019, accessed July 26, 2019. https://www.nationalgeorgarphic.com/animals/2019/07/rare-footage-three-female-great-white-sharks.

7. McField, *Report Card 2010*, 11.

8. Zachos, "Why Are."

9. Jenny Howard, "Why Shark Attacks Are More Common in the Atlantic than the Pacific," *National Geographic*. Last modified July 2, 2019, accessed July 12, 2019. https://www.nationalgeographic.com/animals/2019/07/shark-attacks-atlantic-ocean/.

10. *National Geographic*, "100 Million Sharks Killed Every Year, Study Shows on Eve of International Conference on Shark Protection." Posted March 1, 2013, accessed June 29, 2020. https://www.nationalgeographic.com/culture/onward/2013/03/01/100-million-sharks-killed-every-year-study-shows-on-eve-of-international-conference-on-shark-protection/.

11. Florida Museum of Natural History, International Shark Attack File, "Yearly Worldwide Shark Attack Summary." Accessed July 16, 2019." https://www.floridamuseum.ufl.edu/shark-attacks/yearly-worldwide-summary.

12. Florida Museum of Natural History, "Shark Attack Numbers Remained 'Extremely Low' in 2020, but Fatalities Spiked." Posted January 25, 2021, accessed February 19, 2021. https://www.floridamuseum.ufl.edu/science/shark-attacks-extremely-low-in-2020-fatalities-spiked/

13. Howard, "Why Shark."

14. Zachos, "Why Are."

15. Howard, "Why Shark."

16. Howard, "Why Shark."

17. Florida Museum of Natural History, International Shark Attack File, "Yearly Worldwide."

18. Knowlton, *Citizens*, 161.

19. Florida Museum of Natural History, "Discover Fishes: Shark Biology." Accessed July 13, 2019. https://www.floridamuseum.ufl.edu/discover-fish/sharks/shark-biology/.

20. Florida Museum of Natural History, "Discover Fishes."

21. Florida Museum of Natural History, "Discover Fishes."

22. National Geographic, "Tiger Shark." Accessed July 16, 2019. https://www.nationalgeographic.com/animals/fish/t/tiger-shark/.

23. Howard, "Why Shark."

24. National Geographic, "Tiger Shark."

25. Florida Museum of Natural History, "Discover Fishes."

26. Frank Swain, "World's Fastest Shark Gets a Burst of Speed from Shape-Shifting Skin," *New Scientist.* Last modified March 4, 2019, accessed July 14, 2019. https://www.newscientist.com/article/2195435-worlds-fastest-shark-gets-a-burst-of-speed-from-shape-shifting-skin/.

27. Florida Museum of Natural History, "Discover Fishes."

28. Rutger Thole, "Five Shark Species Your Will Encounter Scuba Diving in the Caribbean Sea." Accessed July 14, 2019. https://rushkult.com/eng/scubamagazine/shark-species-found-in-the-caribbean-sea/.

29. Thole, "Five Shark."

30. William G. Bird, "Sharks," in *Coral Reefs: A Guide to the Common Invertebrates and Fishes of Bermuda, the Bahamas, Southern Florida, the West Indies, and the Coast of Central and South America,* ed. Eugene H. Kaplan (Norwalk, CT: The Easton Press, 1982), 240.

31. Thole, "Five Shark."

32. Florida Museum of Natural History, "Discover Fishes."

33. Bird, "Sharks," 236.

34. Florida Museum of Natural History, "Discover Fishes."

35. National Geographic, "Hammerhead Sharks." Accessed July 17, 2019. https://www.nationalgeographic.com/animals/fish/group/hammerhead-sharks/.

36. Florida Museum of Natural History, "Discover Fishes."

37. National Geographic, "Hammerhead Sharks."

38. National Geographic, "Hammerhead Sharks."

39. Zachos, "Why Are."

40. Florida Museum of Natural History, "Discover Fishes."

41. National Geographic, "Hammerhead Sharks."

42. Stephen Gittings, written communication, October 7, 2019

43. Florida Museum of Natural History, "Discover Fishes."

44. Bird, "Sharks," 236.

45. Florida Museum of Natural History, "Discover Fishes."

46. Bird, "Sharks," 236.

47. Florida Museum of Natural History, "Discover Fishes."

48. Elizabeth Armstrong Moore, "Many Sharks Live a Century–Longer than Thought," *National Geographic.* Posted November 13, 2017, accessed June 8, 2020.

https://www.nationalgeographic.com/news/2017/11/sharks-age-longevity-lifespan-oceans/.

49. Moore, "Many Sharks."

50. Florida Museum of Natural History, "Discover Fishes."

51. Knowlton, *Citizens,* 161.

52. Knowlton, *Citizens,* 161.

53. Florida Museum of Natural History, "Discover Fishes."

54. Florida Museum of Natural History, "Discover Fishes."

55. Florida Museum of Natural History, "Discover Fishes."

56. Florida Museum of Natural History, "Discover Fishes."

57. Bryan Nelson, "12 Animals with the Longest Gestation Period," Mother Nature Network. Last modified April 28, 2010, accessed July 13, 2019. https://www.mnn.com/earth-matters/animals/photos/12-animals-with-the-longest-gestation-period/sharks.

58. Florida Museum of Natural History, "Discover Fishes."

59. Florida Museum of Natural History, "Discover Fishes."

60. Florida Museum of Natural History, "Discover Fishes."

61. Nelson, "12 Animals."

62. World Wildlife Fund, "Mesoamerican Barrier Reef." Last modified April 1, 2012, accessed March 9, 2019. https://www.worldwildlife.org/places/mesoamerican-reef.

63. Castro and Huber, *Marine Biology,* 156.

64. National Geographic, "Whale Shark." Accessed June 8, 2020. https://www.nationalgeographic.com/animals/fish/w/whale-shark/.

65. Knowlton, *Citizens,* 21.

66. World Wildlife Fund, "Mesoamerican."

67. National Geographic, "Whale Shark."

68. National Geographic, "Whale Shark."

69. Kenneth Brower, "Meso Amazing," *National Geographic.* Last modified October 2012, accessed March 9, 2019. https://www.nationalgoegraphic.com/magazine/2012.10/mesoamerican-reef/, 431.

Chapter 15. Invasion of the Lionfish

1. National Oceanic and Atmospheric Administration website. Last modified January 15, 2016, accessed December 5, 2016. https://www.noaa.gov.

2. U.S. Geological Survey, "Taming the Lion(fish)." Posted July 27, 2016, accessed October 26, 2020. http://usgs.gov/news/taming-lionfish.

3. U.S. Geological Survey, "Taming."

4. Fabio Moretzsohn, J. A. Sánchez Chávez, and John Wesley Turnell Jr., eds., "Invasive Species," Gulf Base Resources Database for Gulf of Mexico Research (Corpus Christi: Harte Research Institute for Gulf Studies, Texas A&M University-Corpus Christi, 2004).

5. Jeff MacGregor, "Taming the Lionfish," *Smithsonian,* June 2018, 28.

6. Lad Akins, "Lionfish: Managing the Invasion," *Alert Diver 32,* no. 2 (Spring 2016), 22.

7. Jonathan Peake et al., "Feeding Ecology of Invasive Lionfish (*Pterois volitans* and *Pterois miles*) in the Temperate and Tropical Western Atlantic," *Biological Invasions*. Posted April 11, 2018, accessed January 21, 2019. https://doi.org/10.1007/s10530-018-1720-5.

8. MacGregor, "Taming," 30.

9. MacGregor, "Taming," 35.

10. U.S. Geological Survey, "Nonindigenous Aquatic Species." Last modified June 20, 2016, accessed January 3, 2019. https:/nas.er.usgs.gov.

11. McField, *Report Card 2010*, 10.

12. McField, *Report Card 2010*, 10.

13. Divers Alert Network (D.A.N.), "First Aid Steps after a Lionfish Sting." Last modified August 1, 2014, accessed September 3, 2017. http://www.scubadiving.com/article/news/lionfish-safety-what-to-do-if-stung.

14. Jennifer Chapman, phone interview, November 16, 2016.

15. Florence Fabricant, "Eradicating Invasive Species One Sushi Roll at a Time," *New York Times*. Last modified April 19, 2016, accessed September 2, 2017. http://www.nytimes.com/2016/04/20/dining/invasive.

16. Bun Lai, "How (and Why) to Eat Invasive Species," also published as "Invasive Species Menu of a World-Famous Chef." Last modified September 1, 2013, accessed September 2, 2017. http://www.scientificamerican.com/article/invasive-species-menu-of-a-world-famous-chef.

17. James Morris, Andrew Rhyne, Amber Thomas, and Nancy Breen, "Nutritional Properties of the Invasive Lionfish: A Delicious and Nutritious Approach for Controlling the Invasion," *AACL Bioflux 4*, no. 1, April 2011.

18. Chapman, phone interview, November 16, 2016.

19. MacGregor, "Taming," 36.

20. Monterey Bay Aquarium, "Seafood Watch." Posted 2020, accessed June 30, 2020. https://www.seafoodwatch.org/seafood-recommendations/groups/lionfish.

21. Chapman, phone interview, November 16, 2016.

22. Chapman, phone interview, November 16, 2016.

23. Erick J. Núñez-Vázquez, Antonio Almazán-Becerril, Cavid J. López-Córtes, Alejandra Heredia-Tapia, Francisco E. Hernández-Sandoval, Christine J. Band-Schmidt, José J. Bustillos-Guzmán, Ismael Gárate-Lizárraga, Ernesto García-Mendoza, Cesar A. Salinas-Zavala, Amaury Cordero-Tapia, "Ciguatera in Mexico (1984–2013)," *Marine Drugs 17*, no. 1, December 28, 2018.

24. Chapman, phone interview, November 16, 2016.

25. Chapman, phone interview, November 16, 2016.

26. Chapman, phone interview, November 16, 2016.

27. Dorian Hobday, Priyanka Chadha, Asmat H. Din, and Jenny Geh, "Denaturing the Lionfish," *Eplasty 16*, no. 20, 2016. Posted May 16, 2016, accessed June 30, 2020. https://www.ncbi.nlm.nih.gov/pmc/articles/PMC4892334/.

28. Chapman, phone interview, November 16, 2016.

29. McField, *Report Card 2010*, 10.

30. Jim Waymer, "Lionfish vs. Moray Eel: No Contest," *Florida Today*. Last modified February 24, 2017, September 4, 2017. https://www.floridatoday.com/story/news/local/environment/2017.

31. Ian Drysdale, phone interview, October 20, 2016.

32. Lad Akins, "Lionfish: Managing the Invasion," *Alert Diver 32*, no. 2, Spring 2016, 22.

33. Atkins, "Lionfish," 22.

34. MacGregor, "Taming," 29.

35. Becca Hurley, "Divers Remove 19,167 Lionfish during World's Largest Lionfish Tournament in Florida," *SportDiver*. Posted May 23, 2019, accessed January 22, 2020. https://www.sportdiver.com/emerald-coast-lionfish-tournament.

36. MacGregor, "Taming," 37.

37. MacGregor, "Taming," 35.

38. National Oceanic and Atmospheric Administration, "Impact of Invasive Lionfish." Last updated March 30, 2020, accessed June 30, 2020. https://www.fisheries.noaa.gov/southeast/ecosystems/impacts-invasive-lionfish.

39. Akins, "Lionfish," 21.

40. Atkins, "Lionfish," 20.

41. McField, *Report Card 2010*, 10.

42. Akins, "Lionfish," 21.

43. MacGregor, "Taming," 35.

44. MacGregor, "Taming," 35.

45. Moses Kairo, Bibi Ali, Oliver Cheesman, Karen Haysom, and Sean Murphy, "Invasive Species Threats to the Caribbean: Report to the Nature Conservancy." Last modified 2020, accessed June 10, 2020. http://www.issg.org/database/species/references_files/kairo.

46. National Oceanic and Atmospheric Administration, "Invasive Cup Coral," Flower Gardens Banks National Marine Sanctuary. Last modified July 31, 2017, accessed September 3, 2017. https://flowergardens.noaa.gov/invasivecupcoral.html.

47. National Oceanic and Atmospheric Administration, "Invasive."

48. Canal de Panamá, "The Expanded Canal." Last modified 2017, accessed September 3, 2017. https://micanaldepanama.com/expansion.

Chapter 16. The View from Carrie Bow Cay: The Importance of Monitoring

1. Karen Koltes, in-person interview, September 11, 2016.

2. Koltes, in-person interview, September 11, 2016.

3. Klaus Rützler and Ian G. Macintyre, "The Habitat Distribution and Community Structure of the Barrier Reef Complex at Carrie Bow Cay, Belize," in Smithsonian Contributions to the Marine Sciences, no. 12, *The Atlantic Barrier Reef Ecosystem at Carrie Bow Cay, Belize, I: Structure and Communities*, eds. Klaus Rützler and Ian G. Macintyre (Washington, DC: Smithsonian Institution Press, 1982), 9.

4. Rützler and Macintyre, "Habitat," 44.

5. Koltes, written communication, October 7, 2016.

6. Koltes, written communication, October 13, 2016.

7. Koltes, written communication, October 13, 2016.

8. Koltes, in-person interview, September 11, 2016.

9. Koltes, in-person interview, September 11, 2016.

10. Koltes, written communication, October 7, 2016

11. Koltes, in-person interview, September 11, 2016.

12. Koltes, in-person interview, September 14, 2016.

13. McField et al., *Report Card 2020*, 6.

14. *The Guardian*, "Tropical Storm Iota May Bring More Damage to Caribbean after Eta." Posted November 14, 2020, accessed November 14, 2020. http://theguardian.com/world/2020/nov/14/tropical-storm-iota-may-bring-more-damage-to-caribbean-after-eta.

15. Matthew Cappucci and Jason Samenow, "Hurricane Delta Erupts to Category 4 Strength as It Targets Cancun. Louisiana to Be Hit Later This Week," *Washington Post*, October 6, 2020. http://washingtonpost.com/weather/2020/10/06/hurricane-delta-cancun-louisiana/.

16. CNN Editorial Research, "2020 Atlantic Hurricane Season Fast Facts." Posted October 29. 2020, accessed November 2, 2020. http://cnn.com/2020/05/11/us/2020-atlantic-hurricane-season-fast-facts/index.html.

17. Michael Guy, Hollie Silverman, and Judson Jones, "Hurricane Eta Rapidly Intensified and Is Now a Major Category 4 Hurricane Nearing Landfall," *CNN*. Posted and accessed November 2, 2020. http://cnn.com/2020/1/02/weather/tropical-storm-eta-Monday/index.html.

18. *The Guardian*, "Tropical Storm Iota."

19. Colin Dwyer, "Hurricane Iota, Weakened but Dangerous, Slams an Already Sodden Central America," National Public Radio. Posted November 17, 2020, accessed November 18, 2020. http://npr.org/22020/11/17/935677971/hurricane-iota-weakened-but-dangerous-slams-an-already-sodden-central-america/.

20. Gress et al., "The Mesoamerican," p. 73.

21. Geophysical Fluid Dynamics Laboratory, National Oceanic and Atmospheric Administration, "Global Warming and Hurricanes." Last modified September 23, 2020, accessed November 2, 2020. http://gfdl.noaa.gov/global-warming-and-hurricanes/.

22. Ian Drysdale, written communication, November 2, 2020.

23. Tschirky, in-person interview, September 8, 2016.

24. Tschirky, in-person interview, September 8, 2016.

25. Koltes, in-person interview, September 11, 2016.

26. Koltes, written communication, October 26, 2016.

27. Koltes, written communication, October 26, 2016.

28. Koltes, written communication, October 26, 2016.

29. Joanna Walczak, in-person interview, September 10, 2016.

30. Klaus Rützler and Ian G. Macintyre, eds., *The Atlantic Barrier Reef Ecosystem at Carrie Bow Cay, Belize, I: Structure and Communities*, Smithsonian Contributions to the Marine Sciences, no. 12 (Washington: DC: Smithsonian Institution Press, 1982).

31. Omar Vidal, phone interview, February 6, 2020.

32. Craig Sherwood, in-person interview, September 14, 2016.

Chapter 17. The Lobstermen's Dilemma: A Model Solution

1. David Nuñez, "Mexico—Quintana Roo—The Vigia Chico Fishing Cooperative," The EcoTipping Points Project: Models for Success in a Time of Crisis Last modified November 2006, accessed July 31, 2019. http://www.ecotippingpoints.org/our-stories/indepth/mexico-quintana-roo-vigia-chico.html.

2. Erica Cunningham, "Catch Shares in Action: Mexican Vigia Chico Cooperative Spiny Lobster Territorial Use Rights for Fishing Program," Environmental Defense Fund, 2013. Accessed July 1, 2020. http://fisherysolutionscenter.edf.org/sites/catch-shares.edf.org/files/Mexican_Vigia_Chico_Spiny_Lobster_TURF.pdf.

3. E. Sosa-Cordero, M.L.A. Liceago-Correra, and J. C. Seijo, "The Punta Allen Lobster Fishery: Current Status and Recent Trends," FAO Fisheries, 2008, 149.

4. Sosa-Cordero et al., "The Punta Allen," 150.

5. Ley-Cooper, *Sustainability*, 83.

6. Nuñez, "Mexico."

7. Ley-Cooper, *Sustainability*, 181.

8. Cristol Méndez-Medina, Brijit Schmook, and Susannah R. McCandless, "The Punta Allen Cooperative as an Emblematic Example of a Sustainable Small-Scale Fishery in the Mexican Caribbean," *Maritime Studies 14*, no. 12, 2015. Posted June 23, 2015, accessed July 1, 2020. https://doi.org/10.1186/s40152-015-0026-9.

9. Méndez-Medina et al., "The Punta Allen Cooperative."

10. Sosa-Cordero et al., "The Punta Allen," 155–156.

11. Ley-Cooper, *Sustainability*, 236.

12. Nuñez, "Mexico."

13. Ley-Cooper, *Sustainability*, 123.

14. World Commission on Environment and Development, *Our Common Future, from One Earth to One World*, March 23, 1987, 3.27.

15. Nuñez, "Mexico."

16. Sosa-Cordero et al., "The Punta Allen," 156.

17. Martha E. Fonseca-Larios and Patricia Briones-Fourzan, "Fecundity of the Spiny Lobster *Panulirus argus* (Latreille, 1804) in the Caribbean Coast of Mexico," *Bulletin of Marine Science-Miami 63*, no. 1, July 1998, 25–27.

18. Ley-Cooper, *Sustainability*, 84.

19. Ley-Cooper, *Sustainability*, 245.

20. This scenario is an amalgam of descriptions of typical meetings of fishing cooperatives in the Mexican Caribbean.

21. Sosa-Cordero et al., "The Punta Allen," 154.

22. Sosa-Cordero et al., "The Punta Allen," 157.

23. Ley-Cooper, *Sustainability*, 202.

24. Nuñez, "Mexico."

25. Garrett Hardin, "The Tragedy of the Commons," *Science 162*, no. 3859, December 13, 1968, 1243–1248.

26. Ley-Cooper, *Sustainability*, 108.

27. Erik Vance, "Building a Better Lobster Trap," *Scientific American*. Posted December 18, 2013, accessed July 1, 2020. https://www.scientificamerican.com/article/building-a-better-lobster-trap/.

28. Florida Keys Commercial Fishermen's Association, *Fishery Nation*. Posted May 13, 2020, accessed July 1, 2020. https://fisherynation.com/archives/tag/florida-keys-commercial-fishermens-association.

29. Ley-Cooper, *Sustainability*, 271.

30. Benjamin C. Gutzler, Mark J. Butler IV, and Donald C. Behringer, "Casitas: a Location-Dependent Ecological Trap for Juvenile Caribbean Spiny Lobsters, *Panulirus argus*," *ICES Journal of Marine Science 17*, no. suppl_1, July 2015, i177–i184.

31. Butler et al., "Transmission," 173.

32. Ley-Cooper, *Sustainability*, 60.

33. Ley-Cooper, *Sustainability*, 111.

34. Ley-Cooper, *Sustainability*, 146.

35. Ley-Cooper, *Sustainability*, 146.

36. Ley-Cooper, *Sustainability*, 151.

37. Ley-Cooper, *Sustainability*, 150.

38. Vidal, phone interview, February 6, 2020.

39. Vidal, phone interview, February 6, 2020.

40. Ley-Cooper, *Sustainability*, 111.

41. Ley-Cooper, *Sustainability*, 234.

42. Ley-Cooper, *Sustainability*, 165.

43. Ley-Cooper, *Sustainability*, 180.

44. Ley-Cooper, *Sustainability*, 229.

45. Ley-Cooper, *Sustainability*, 230.

46. Ley-Cooper, *Sustainability*, 82.

47. Ley-Cooper, *Sustainability*, 238.

48. Ley-Cooper, *Sustainability*, 239.

49. Ley-Cooper, *Sustainability*, 253.

50. Ley-Cooper, *Sustainability*, 244.

51. Ley-Cooper, *Sustainability*, 128.

52. Ley-Cooper, *Sustainability*, 126.

53. Ley-Cooper, *Sustainability*, 129.

54. Stuart Fulton, phone interview, January 20, 2020.

55. Ley-Cooper, *Sustainability*, 128.

56. Ley-Cooper, *Sustainability*, 235.

57. Ley-Cooper, *Sustainability*, 236.

58. Ley-Cooper, *Sustainability*, 240.

59. Ley-Cooper, *Sustainability*, 236.

Chapter 18. Thinking Globally, Acting Locally—and Collaboratively

1. COBI website, "About." Posted 2016, accessed January 25, 2020. http://cobi.org.mx/en/about-cobi/.

2. David and Lucile Packard Foundation, "Comunidad y Biodiversidad, A.C." Posted 2020, accessed January 25, 2020. https://www.packard.org/grants-and-investments/grants-database/comunidad-y-biodiversidad-a-c/.

3. COBI, "About."

4. Fulton, phone interview, January 20, 2020.

5. Fulton, phone interview, January 20, 2020.

6. Fulton, phone interview, January 20, 2020.

7. Fulton, phone interview, January 20, 2020.

8. Fulton, phone interview, January 20, 2020.

9. Healthy Reefs Initiative, "About Us." Posted 2020, accessed June 17, 2020. https://www.healthyreefs.org/cms/about-us/.

10. Melanie McField, personal communication, February 11, 2021.

11. Healthy Reefs Initiative," Our Team." Posted 2020, accessed June 17, 2020. https://www.healthyreefs.org/cms/our-team.

12. Fulton, phone interview, January 20, 2020.

13. Southern Environmental Association, "Research." Posted 2020, accessed February 7, 2020. https://www.seabelize.org/research/.

14. Rueda Flores, phone interview, November 3, 2016.

15. Melanie McField, *Report Card for the Mesoamerican Reef 2016.* (Playa del Carmen, Mexico: Healthy Reefs for Healthy People Healthy Reefs Initiative, 2016).

16. Index Mundi, "Guatemala Literacy." Accessed August 7, 2019. https://www.indexmundi.com/guatemala/literacy.html.

17. Giró Petersen, phone interview, October 12, 2016.

18. Healthy Reefs Initiative website. Accessed August 7, 2019. https://www.healthyreefs.org.

19. Pott, phone interview, October 12, 2016.

20. Arturo Dominici, Bill Dennison, Chad Englis, Guilherme Dutra, Leah Bruce Karrer, and Pacifico Beldia, *A Decision-Maker's Guide to Using Science/A Scientist's Guide to Influencing Decision-Making* (Arlington, VA: Conservation International, Science and Knowledge Division, 2011), DM-3.

21. Dominici et al., "A Decision-Maker's," DM-4.

22. Dominici et al., "A Decision-Maker's," DM-4.

23. Dominici et al., "A Decision-Maker's," S-4.

24. Dominici et al., "A Decision-Maker's," S-6.

25. Dominici et al., "A Decision-Maker's," S-7.

26. McField, phone interview, October 10, 2016.

27. Wildlife Conservation Society, "Government of Belize Expands Marine Protected Areas in Biodiverse Offshore Waters." Posted April 3, 2019, accessed February 15, 2021. http://newsroom.wcs.org/News-Releases/articleType/ArticleView/12150/

government-of-belize-expands-marine–protected-areas-in-biodiverse-offshore-waters/.

28. Emily Frost, "Improving Grades for the Mesoamerican Barrier Reef," *Smithsonian*. Last modified March 2018, accessed July 22, 2019. https://ocean.si.edu/ocean-life/plants-algae/improving-grades-mesoamerican-reef.

29. McField, "Eco-Audit (2016)."

30. McField, *Report Card 2010*, 16.

31. Giró Petersen, phone interview, October 12, 2016.

32. Gustavo Guerrero, phone interview, January 15, 2020.

33. Guerrero, phone interview, January 15, 2020.

34. Julian Smith, "Bracing for Impact," *Nature Conservancy*, Winter 2018, 50–59.

35. *San Pedro Sun*, "Belize Advances in Climate Change Adaptation Planning and Disaster Risk Management with Support from UB and IDB." Posted November 9, 2019, accessed February 8, 2020. https://www.sanpedrosun.com/government/2019/11/09/belize-advances-in-climate-change-adaptation-planning-and-disaster-risk-management-with-support-from-ub-and-idb.

36. Green Climate Fund, "Belize Country Strategic Framework Including a Country Programme for Engagement with the Green Climate Fund," Final Version, June 2019, 24. Posted August 12, 2019, accessed February 8, 2020. https:/www.acclimatise.uk.com/wp-content/uploads/019/08/BELIZE-CP-Final-draft-version_12.08.19-CLEAN.pdf.

37. Vidal, phone interview, February 6, 2020.

38. Vidal, phone interview, February 6, 2020.

39. Vidal, phone interview, February 6, 2020.

40. Vidal, phone interview, February 6, 2020.

41. World Wildlife Fund, "Mesoamerican Barrier Reef." Last modified April 11, 2012, accessed March 9, 2019. https://www.worldwildlife.org/places/mesoamerican-reef.

42. *Mexico News Daily*, "New Protected Area to Be Mexico's Biggest." Last modified December 3, 2016, accessed August 12, 2019. https://mexiconewsdaily.com/news/new-protected-area-to-be-mexicos-biggest/.

43. Giró Petersen, phone interview, October 12, 2016.

44. Vidal, phone interview, February 6, 2020.

45. Oscar Lopez, "'Coral Bleaching Is Getting Worse . . . but the Biggest Problem is Pollution,'" *The Guardian*. Last modified December 27, 2017, accessed August 13, 2019. https://www.theguardian.com/environment/2017/dec/27/coral-bleaching-is-getting-worse-but-the-biggest-problem-is-pollution.

46. Frost, "Improving."

47. McField, phone interview, October 10, 2016.

48. Les Kaufman and John Tschirky, *Living with the Sea: Local Efforts Buffer Effects of Global Change* (Arlington, VA: Conservation International, Science and Knowledge Division, Arlington, VA, 2010), 5.

49. Frost, "Improving."

50. John Cannon, "Parrotfish, Critical to Reef Health, Now Protected Under Mexican Law," *Mongabay.* Posted November 7, 2018, accessed February 21, 2021. https://news.mongabay.com/2018/11/parrotfish-critical-to-reef-health-now-protected-under-mexican-law/.

51. McField, *Report Card 2010,* 2.

52. World Population Review, "Belize Population." Posted February 2021, accessed February 12, 2021. https://www.worldpopulationreview.com/countries/belize-population.

53. Knoema, "Quintana Roo—Total Population." Posted January 2021, accessed February 14, 2021. http://knoema.com/atlas/Mexico/Quintana-Roo/topics/Demographics/Key-Indicators/Population.

54. Macrotrends, "Guatemala." Posted January 2021, accessed February 12, 2021. https://www.macrotrends.net/countries/GTM/gutemala/population-growth-rate/.

55. Macrotrends, "Honduras." Posted January 2021, accessed February 12, 2021. https://www.macrotrends.net/countries/HON/honduras/population-growth-rate/.

56. *Encyclopaedia Britannica,* "Bay Islands." Accessed August 7, 2019. https://www.britannica.com/place/Bay-Islands.

57. McField, written communication, February 9, 2021.

58. Giró Petersen, phone interview, October 12, 2016.

59. Giró Petersen, phone interview, October 12, 2016.

60. Giró Petersen, phone interview, October 12, 2016.

61. Drysdale, phone interview, October 12, 2016.

62. Drysdale, phone interview, October 12, 2016.

63. Drysdale, phone interview, October 12, 2016.

64. Roatán Online, "Roatán Marine Park." Posted 2020, accessed July 2, 2020. https://roatan.online/roatan-marine-park.

65. Drysdale, phone interview, October 12, 2016.

66. Pott, phone interview, October 12, 2016.

67. Summit Foundation, "Clean Water for Clean Reefs in Honduras," July 11, 2017 and grants report 2020. https://www.summitfdn.org/mesoamerican-reef/clean-water-for-clean-reefs-in-honduras/.

68. Coral Reef Alliance (CORAL), "Coral Reefs in Roatán Thrive with Clean Water." Posted 2020, accessed July 3, 2020. http://coral.org/blog/coral-reefs-in-roatan-thrive-with-clean-water/.

69. Drysdale, phone interview, October 12, 2016.

70. Frost, "Improving."

71. McField, *Report Card 2010.*

72. Jennifer Myton, phone interview, November 10, 2016.

73. Kaufman and Tschirky, *Living,* 6.

74. Myton, phone interview, November 10, 2016.

75. Myton, phone interview, November 10, 2016.

76. Pott, phone interview, October 12, 2016.

77. Andrea Kealoha, phone interview, January 15, 2020.

78. Kealoha, phone interview, January 15, 2020.

79. McField, *Report Card 2016.*

80. Pro México, "Mexico's Seafood Industry." Accessed January 22, 2020. http://www.gob.mx/cms/uploads/attachment/file/637431/seafood-industry.pdf.

81. Marisol Rueda Flores, phone interview, November 3, 2016.

82. Rueda Flores, phone interview, November 3, 2016.

83. Michael Orbach and Leah Karrer, *Marine Managed Areas: What, Why, and Where* (Arlington, VA: Science and Knowledge Division, Conservation International, 2010), 4.

84. Orbach and Karrer, "Marine," 3.

85. Kaufman and Tschirky, *Living,* 5.

86. Kaufman and Tschirky, *Living,* 3.

87. Giselle Samonte, Leah Bunce Karrer, and Michael Orbach, *People and Oceans: Managing Marine Areas for Human Well-Being* (Arlington, VA: Science and Knowledge Division, Conservation International, 2010), 10.

88. McField, *Report Card 2016.*

89. Pott, phone interview, October 12, 2016.

90. Samonte et al., *People,* 4.

91. Island Expeditions, "Lighthouse Reef: Belize's First Marine Protected Area." Accessed August 11, 2019. https://www.islandexpeditions.com/belize-vacations-blog/lighthouse-reef-belizes-first-marine-protected-area.

92. Laughing Bird, "Welcome to Laughing Bird Caye National Park in Belize." Last modified 2019, accessed August 11, 2019. https://www.laughingbird.org/park1.html.

93. Pott, phone interview, October 12, 2016.

94. Vidal, phone interview, February 6, 2020.

95. McField, phone interview, October 10, 2016.

96. McField, phone interview, October 10, 2016.

97. Kaufman and Tschirky, *Living,* 6.

98. McField, *Report Card 2010,* 14.

99. McField, phone interview, October 10, 2016.

100. Nicole Helgason, "Coral Bleaching Hits Mesoamerican Reef." Last modified October 17, 2017, accessed July 22, 2019. https://reefdivers.io/coral-bleaching-hits-hard/4666.

101. Elizabeth Kolbert, "Unnatural Selection: What Will It Take to Save the World's Reefs and Forests?" *New Yorker,* April 18, 2016, 22.

102. McField, *Report Card 2010,* 14.

103. Helgason, "Coral Bleaching."

104. McField, *Report Card 2010,* 17.

105. McField, *Report Card 2010,* 18.

106. Drysdale, phone interview, October 12, 2016.

107. McField, "Healthy Reefs for Healthy People Eco Audit (2016)."

108. McField, phone interview, October 10, 2016.

109. McField, "Healthy . . . (2016)."

110. McField, phone interview, October 10, 2016.

111. Courtney Leatherman, "Interview: Stephanie Wear," *Nature Conservancy,* March-April 2013, 27.

112. Pott, phone interview, October 12, 2016.

113. Chapman, phone interview, November 16, 2016.

114. Chapman, phone interview, November 16, 2016.

115. Giró Petersen, phone interview, October 12, 2016.

116. *New York Times,* obituary, "Ruth Gates." Last modified November 5, 2018, accessed August 14, 2019. https://www.nytimes.com/2018/11/05/obituaries/ruth-gates-dead-marine-biologist-who-championed-coral.html.

117. Kolbert, "Unnatural," 24.

118. Kolbert, "Unnatural," 23.

119. Kolbert, "Unnatural," 23.

120. Matt Dozier, "The Restoration Revolution: Gearing Up for the Largest Coral Restoration Project Ever in the Florida Keys," *Alert Diver 33,* no. 1, Winter 2017, 18.

121. Dozier, "Restoration," 19.

122. McField, *Report Card 2010,* 18.

123. Nuno Simoes, phone interview, January 21, 2020.

124. Simoes, phone interview, January 21, 2020.

125. Simoes, phone interview, January 21, 2020.

126. Myton, phone interview, November 3, 2016.

Chapter 19. And Individually: What Each of Us Can Do Right Now to Save the Mesoamerican Barrier Reef

1. Carolyn Kormann, "The Widening Gyre: A Young Entrepreneur Takes on Ocean Pollution," *New Yorker,* February 4, 2019, 42.

2. Kormann, "The Widening," 44.

3. Kormann, "The Widening," 45.

4. Kormann, "The Widening," 44.

5. Boyan Slat, "Into the Twilight Zone," The Ocean Cleanup. Posted August 16, 2019, accessed August 16, 2019. https://theoceancleanup.com/updates/.

6. Ocean Cleanup, "The Ocean Cleanup Successfully Catches Plastic in the Great Pacific Garbage Patch." Posted October 2, 2019, accessed February 23, 2021. https://theoceancleanup.com/updates/the-ocean-cleanup-successfully-catches-plastic-in-the-great-pacific-garbage-patch/.

7. Kormann, "The Widening," 48.

8. Ocean Cleanup, "Interceptor 011 in Place after Extreme Flooding in Jakarta." Posted January 22, 2020, accessed February 23, 2021. https://theoceancleanup/general/interceptor-011-in-place-after-extreme-flooding-in-jakarta/.

9. Ocean Cleanup, "Thailand Interceptor Project Kicks Off with Signing of MOU." Posted February 3, 2021, accessed February 23, 2021. https://theoceancleanup.com/updates/thailand-interceptor-project-kicks-off-with-signing-of-mou/.

10. Ocean Cleanup, "Turning Trash into Treasure: The Ocean Cleanup Sunglasses." Posted October 24, 2020, accessed February 23, 2021. https://theoceancleanup.com/general/turning-trash-into-treasure-the-ocean-cleanup-sunglasses/.

11. National Oceanic and Atmospheric Administration, Ocean Facts, "What Is a Gyre?" Accessed August 16, 2019. https://oceanservice.noaa.gov/facts/gyre.html.

12. Encyclopaedia Britannica online, "Industrial Revolution." Last modified September 4, 2019, accessed June 12, 2020. https://www.britannica.com/event/industrial-revolution.

13. Michelle Z. Donahue, "Dino-Killing Asteroid Hit Just the Right Spot to Trigger Extinction," *National Geographic.* Posted November 9, 2017, accessed August 18, 2019. https://www.nationalgeographic.com/news/2017/11/dinosaurs-extinction-asteroid-chicxulub-soot-earth-science/.

14. Marine Stewardship Council, "Sustainable Fish to Eat." Accessed August 18, 2019. https://www.msc.org/what-you-can-do/eat-sustainable-seafood/fish-to-eat.

15. Oceana, "Sustainable Seafood Guide." Accessed August 18, 2019. https://oceana.org/living-blue/sustainable-seafood-guide.

16. Myton, phone interview, November 3, 2016.

17. Craig Downs, "Sunscreen Pollution: A Serious and Increasingly Clear Threat to Coral," *Alert Diver online,* Winter 2016. Accessed August 20, 2019. http://www.alert-diver.com/Sunscreen-Pollution.

18. Will Coldwell, "Hawaii Becomes First US State to Ban Sunscreens Harmful to Coral Reefs," *The Guardian.* Last modified May 3, 2018, accessed August 20, 2019. https://www.theguardian.com/travel/2018/may/03/hawaii-becomes-first-us-state-to-ban-sunscreens-harmful-to-coral-reefs.

19. Downs, "Sunscreen."

20. Downs, "Sunscreen."

21. Matthew S. Schwartz, "Key West to Ban Popular Sunscreen Ingredients to Protect Coral Reef," National Public Radio. Last modified February 6, 2019, accessed August 20, 2019. https://www.npr.org/2019/02/06/691913378/key-west-votes-to-ban-popular-sunscreen-ingredients-to-protect-coral-reef.

22. Ana Ceballos, "Ban Certain Sunscreens? Florida Senate Committee Blocks Cities, Counties from Doing It," *Orlando Sentinel.* Last modified April 9, 2019, accessed August 19, 2019. https://www.orlandosentinel.com/business/os-bz-florida-no-sunscreen-ban-20190409-story.html.

23. Coldwell, "Hawaii."

24. Kevin Mallory, in-person interview, February 2, 2019.

25. Downs, "Sunscreen."

26. Amy Marturana, "This Is What Actually Happens to Your Skin When You Get a Tan," *Self.* Posted June 8, 2016, accessed August 20, 2019. https://www.self.com/story/this-is-what-actually-happens-to-your-body-when-you-get-a-tan.

27. Rueda Flores, telephone interview, November 3, 2016.

28. Vidal, phone interview, February 6, 2020.

29. Cruise Lines International Association, "Will the CDC Lift the 'No Sail'

Order?" *Cruise Industry News.* Posted October 26, 2020, accessed October 29, 2020. http://cruiseindustrynews.com/cruise-news/23753/wil-the-cdc-lift-the-no-sail-order.html.

30. Sarah Mervosh, "Carnival Cruises to Pay $20 Million in Pollution and Cover-Up Case," *New York Times.* Posted June 4, 2019, accessed October 30, 2020. http://nytimes.com/2019/06/04/businesss/carnival-cruise-pollution.html.

31. Mervosh, "Carnival Cruises."

32. Mervosh, "Carnival Cruises."

33. Cruise Lines International, "Will the CDC."

34. Cruise Lines International, "Will the CDC."

35. Cruise Critic, "When Are Cruise Lines around the World Expected to Resume Service?" Posted February 22, 2021, accessed February 24, 2021. htps://www.cruise-critic.com?news/5206/.

36. Ceylan Yeginsu, "Where Cruise Ships Are Sent to Die," *New York Times,* October 30, 2020. Accessed October 30, 2020. http://nytimes.com/2020/10/30/travel/cruise-ships-scrapped.html.

37. Nina Burleigh, "The Caribbean Dilemma," *New York Times.* Posted August 4, 2020, accessed October 29, 2020. http://nytimes.com/2020/08/04/travel/coronavirus-caribbean-vacations.

38. Burleigh, "The Caribbean."

39. Ian Drysdale, written communication, November 2, 2020.

40. Drysdale, written communication, November 2, 2020.

BIBLIOGRAPHY

Akins, Lad. "Lionfish: Managing the Invasion." *Alert Diver 32*, no. 2 (Spring 2016): 20–22.

Ambergriscaye.com. "The Great Blue Hole at Lighthouse Reef." Accessed August 31, 2017. https://ambergriscaye.com/pages/town/greatbluehole.html.

Ambler. "Meet the Aquatic Species of the Great Blue Hole in Belize." Accessed August 30, 2017. http://www.amble.com/ambler/species-of-the-great-blue-hole-in-belize.

Anderson, Roland C., Jennifer A. Martin, and James B. Wood. *Octopus: The Ocean's Intelligent Invertebrate*. Portland, OR: Timber Press, 2013.

Answers.com. Accessed February 6, 2019. https://www.answers.com.

Arnason, Ulfur, Annette Gullberg, and Axel Janke. "Molecular Timing of Primate Divergences as Estimated by Two Nonprimate Calibration Points." *Journal of Molecular Evolution 37*, no. 6 (December 1998): 718–727.

Associated Press. "Ray Leaps from Water, Killing a Boater in the Florida Keys." Last modified March 21, 2008, accessed July 14, 2019. https://www.nytimes.com/2008/03/21/us/21sting.html.

Barneche, Diego, and Dustin Marshall. "Fishy Limits." *Alert Diver 35*, no. 1 (Winter 2019): 110–112.

Belize Audubon Society. "Half Moon Caye Natural Monument." Last modified 2016, accessed March 12, 2019. https://www.belizeaudubon.org.

Belize.com. "Demographics." Accessed March 12, 2019. https://belize.com/demographics/.

———. "History of the Garifuna People." Accessed March 12, 2019. https://belize,com/history-of-the-garifuna-people.

———. "Turneffe Atoll." Accessed January 7, 2019. https://belize.com/turneffe-atoll.

Berkeley Evolibrary. "The Evolution of Whales." Accessed July 7, 2019. https://evolution.berkeley.edu/evolibrary/article/evograms_03.

Berwald, Juli. *Spineless: The Science of Jellyfish and the Art of Growing a Backbone*. New York: Riverhead Books, 2017.

Biernbaum, Charles K. "Invertebrate Experiments and Research Projects." Accessed January 20, 2019. https://www.gricemarinelab.cofc.edu.

Bird, William G. "Sharks." In *Coral Reefs: A Guide to the Common Invertebrates and Fishes of Bermuda, the Bahamas, Southern Florida, the West Indies, and the Coast of*

Central and South America, edited by Eugene H. Kaplan. Norwalk, CT: The Easton Press, 1982, 226–241.

Bland, Lucie M., Tracey J. Regan, Minh Ngoc Dinh, Renata Ferrari, David A. Keith, Rebecca Lester, David Mouillot, Nicholas J. Murray, Houng Anh Nguyen, and Emily Nicholson. "Meso-American Reef: Using Multiple Lines of Evidence to Assess the Risk of Ecosystem Collapse." *Proceedings of the Royal Society B 28,* no. 183. Last modified September 20, 2017, accessed March 10, 2019. https://dol.org/10.1098/rsph.2017.0660.

Block, Melissa. "*Stenorhynchus seticornis:* Yellowline Arrow Crab." University of Michigan Museum of Zoology, Animal Diversity Web. Accessed July 3, 2019. https://www.animaldiversity.org/accounts/Stenorhynchus_seticornis/.

Blue Planet II. United Kingdom: BBC Natural History Unit, 2017.

Braütigam, Amie, and Karen L. Eckert. *Turning the Tide: Exploitation, Trade and Management of Marine Turtles in the Lesser Antilles, Central America, Colombia and Venezuela.* Cambridge, UK: TRAFFIC: The Wildlife Trade Monitoring Network and CITES Secretariat, 2006.

Bray, Richard T., and Gordon T. Taylor. "Sea Lilies or Feather Stars." In *Coral Reefs: A Guide to the Common Invertebrates and Fishes of Bermuda, the Bahamas, Southern Florida, the West Indies, and the Caribbean Coast of Central and South America,* edited by Eugene H. Kaplan. Norwalk, CT: The Easton Press, 1982, 202–205.

Brevard Times. "Giant Mantis Shrimp Caught off Florida Dock." Last modified November 25, 2015, accessed July 4, 2019. http://news.brevardtimes.com/2014/09/giant-mantis-shrimp-caught-off-florida.html.

Brower, Kenneth. "Meso Amazing." *National Geographic.* Last modified October 2012, accessed March 9, 2019. https://www.nationalgoegraphic.com/magazine/2012.10/mesoamerican-reef/.

Brown, Desmond. "Caribbean Fears Loss of 'Keystone Species' to Climate Change." Inter Press Service. Last modified April 2014, accessed March 22, 2019. http://www.ipsnews.net/2014/04/caribbean-fears-loss-keystone-species-climate-change/.

Bulfinch, Thomas. *Bulfinch's Mythology.* London: Spring Books, 1964.

Burgess, Warren E. *Corals.* Neptune, NJ: T.F.H. Publications, 1979.

Burleigh, Nina. "The Caribbean Dilemma." *New York Times.* Posted August 4, 2020, accessed October 29, 2020. http://nytimes.com/2020/08/04/travel/coronavirus-caribbean-vacations.

Canal de Panamá. "The Expanded Canal." Last modified 2017. https://micanaldepanama.com/expansion.

Canisius Ambassadors for Conservation. "Clownfish." Accessed February 25, 2019. http://www.conservenature.org/learn_about_wildlife/great_barrier_reef/clownfish.htm.

Cannon, John, "Parrotfish, Critical to Reef Health, Now Protected Under Mexican Law." *Mongabay.* Posted November 7, 2018, accessed February 21, 2021. http://www.news.mongabay.com/parrotfish-critical-to-reef-health-now-protected-unhder-mexican-law/.

Cappucci, Matthew, and Jason Samenow. "Hurricane Delta Erupts to Category 4 Strength as It Targets Cancun. Louisiana to Be Hit Later This Week." *Washington Post,* October 6, 2020. http://washingtonpost.com/weather/2020/10/06/hurricane-delta-cancun-louisiana/.

Carlton, Michael. "Ixtapa: Mexico's Gem of a Resort." *Spokesman-Review,* January 24, 1982, B6.

Carson, Rachel. *Silent Spring.* Boston: Houghton Mifflin, 1962.

Castro, Joseph. "How Do Dolphins Sleep?" *Live Science.* Last modified April 14, 2014, accessed July 8, 2019. https://www.livescience.com/how-do-dolphins-sleep/html.

Ceballos, Ana. "Ban Certain Sunscreens? Florida Senate Committee Blocks Cities, Counties from Doing It." *Orlando Sentinel.* Last modified April 9, 2019, accessed August 19, 2019. https://www.orlandosentinel.com/business/os-bz-florida-no-sunscreen-ban-20190409-story.html.

Central Intelligence Agency. "Poverty in Guatemala." Accessed February 24, 2019. https://www.cia.gov/library/publications/the-world-facebook/geos/print-gt.html.

Chepkemoi, Joy. "Where Is the Mesoamerican Barrier Reef Located?" WorldAtlas. Last modified April 25, 2017, accessed March 9, 2019. https://www.worldatlas.com/articles/where-is-the-mesoamerican-barrier-reef-system-located.html.

CNN Editorial Research. "2020 Atlantic Hurricane Season Fast Facts." Posted October 29. 2020, accessed November 2, 2020. http://cnn.com/2020/05/11/us/2020-atlantic-hurricane-season-fast-facts/index.html.

COBI. "History." Posted 2016, accessed January 25, 2020. http://cobi.org.mx/en/about-cobi/history.

Coldwell, Will. "Hawaii Becomes First US State to Ban Sunscreens Harmful to Coral Reefs." *The Guardian.* Last modified May 3, 2018, accessed August 20, 2019. https://www.theguardian.com/travel/2018/may/03/hawaii-becomes-first-us-state-to-ban-sunscreens-harmful-to-coral-reefs.

Cole, Brandon. "Cozumel Revisited." *Alert Diver 34,* no. 3 (Summer 2018): 78–83.

Countrymeters [website]. Accessed February 6, 2019. https://www.countrymeters.info/en/World.

Cruise Critic, "When Are Cruise Lines around the World Expected to Resume Service?" Posted February 22, 2021, accessed February 24, 2021. htps://www.cruise-critic.com?news/5206/.

Cruise Industry International Association. "Will the CDC Lift the 'No Sail' Order?" *Cruise Industry News.* Posted October 26, 2020, accessed October 28, 2020. http://cruiseindistrynews.com/cruise-news/23758-will-the-cdc-lift-te-no-sail-order.html.

Cruise Lines International Association. Accessed October 28, 2020. http://cruising.org/en/cruise-lines.

Curtis, Vickie. "Chasing Coral." *Alert Diver 33,* no. 4 (Fall 2017): 89–91.

Darwin, Charles. *The Structure and Distribution of Coral Reefs.* Tucson: University of Arizona Press, 1984 (originally published in 1842).

David and Lucile Packard Foundation. "Comunidad y Biodiversidad, A.C." Posted 2020, accessed January 25, 2020. https://www.packard.org/grants-and-invest-ments/grants-database/comunidad-y-biodiversidad-a-c/.

Davis, Josh. "Shark Evolution: A 450 Million Year Timeline." Natural History Museum. Last modified December 13, 2018, accessed July 16, 2019. https://www.nhm.ac.uk/discover/shark-evolution-a-450-million-year-timeline.html.

Davis, Sherri. "Weekly Cruise Ship Arrival and Departure Information." *Cozumel Insider.* Last modified 2019, accessed March 15, 2019. https://cozumelinsider.com/?Page=CruiseShips.

DeLoach, Ned, and Anna DeLoach. "How to Save a Fish." *Alert Diver 34,* no. 3 (Summer 2018): 32–33.

———. "Leaping for Love." *Alert Diver 34,* no. 4 (Fall 2018): 34–35.

———. "Pest Control: Part Two: The Manicure." *Alert Diver 35,* no. 2 (Spring 2019): 32–33.

Detecon Consulting. "Speaking of Starfish and Mussels: Role Distribution and Success Strategies in Business Ecosystems." Accessed March 24, 2019. https://www.detecon.com/en/knwledge/speaking-stafish-and-mussels.

Divers Alert Network (D.A.N.). "First Aid Steps after a Lionfish Sting." Last modified August 1, 2014, accessed September 3, 20117. https://www.scubadiving.com/article/news/lionfish-safety-what-to-do-if-stung.

Dolphin Communication Project [website]. Accessed July 6, 2019. https://www.dolphincommunicationproject.org/index.php/the-latest-buzz/field-reports/bahamas-3/bahamas-2000/item/94230-stan-kuczaj-ph-d.

Dominici, Arturo, Bill Dennison, Chad Englis, Guilherme Dutra, Leah Bunce Karrer, and Pacifico Beldia. *A Decision-Maker's Guide to Using Science/A Scientist's Guide to Influencing Decision-Making.* Arlington, VA: Conservation International, Science and Knowledge Division, 2011.

Donahue, Michelle Z. "Dino-Killing Asteroid Hit Just the Right Spot to Trigger Extinction." *National Geographic.* Last modified November 9, 2017, accessed August 18, 2019. https://www.nationalgeographic.com/news/2017/11/dinosaurs-extinction-asteroid-chicxulub-soot-earth-science/.

Downs, Craig. "Sunscreen Pollution: A Serious and Increasingly Clear Threat to Coral." *Alert Diver Online,* Winter 2016. Accessed August 20, 2019. https://www.alertdiveronline/.

Dozier, Matt. "The Restoration Revolution: Gearing Up for the Largest Coral Restoration Project Ever in the Florida Keys." *Alert Diver 33,* no. 1 (Winter 2017): 18–21.

Dwyer, Colin. "Hurricane Iota, Weakened but Dangerous, Slams an Already Sodden Central America." National Public Radio. Posted November 17, 2020, accessed November 18, 2020. http://npr.org/22020/11/17/935677971/hurricane-iota-weakened-but-dangerous-slams-an-already-sodden-central-america/.

Egner, Sarah. "Differentiating Coral Bleaching and Coral Mortality: A Case Study from the Great Barrier Reef." *Alert Diver 33,* no. 1 (Winter 2017): 102–105.

———. "Parrotfish: Grazers of the Reef." *Alert Diver 32,* no. 3 (Summer 2016): 38–40.

Ellis, Richard. *Singing Whales and Flying Squid: The Discovery of Marine Life*. Guilford, CT: The Lyons Press, 2005.

Encyclopaedia Britannica online. "Bay Islands." Accessed August 7, 2019. https://www.britannica.com/place/Bay-Islands.

Encyclopaedia Britannica online. "Coral Gall Crab." Accessed July 3, 2019. https://www.britannica.com/animals/coral-gall-crab.

Engman, James. "Mass Mortality in *Diadema antillarum* (Echinodermata: Echinoidea): A Large-Scale Natural Experiment in Herbivore Removal." Last modified October 29, 2002, accessed February 9, 2019. https://www.jrscience.wcmuchio.edu.

Eschmeyer, William. *The Catalog of Fishes*. Maintained online by the California Academy of Sciences. Accessed January 29, 2019. https://www.researcharchive.calacademy.org/research/ichthyology.

Fabricant, Florence. "Eradicating Invasive Species One Sushi Roll at a Time." *New York Times*. Last modified April 19, 2016, accessed September 2, 2017. https://www.nytimes.com/2016/04/20/dining/invasive.

Fink, Stephen. "Publisher's column." *Alert Diver 34*, no. 4 (Fall 2018): 10–11.

Florida Fish and Wildlife Conservation Commission. Accessed February 1, 2019. https://myfwc.com/wildlifehabitats/profiles.

Florida Museum of Natural History. International Shark Attack File, "Yearly Worldwide Shark Attack Summary." Accessed July 16, 2019. https://www.floridamuseum.ufl.edu/shark-attacks/yearly-worldwide-summary/.

———. "Discover Fishes: Shark Biology." Accessed July 13, 2019. https://www.floridamuseum.ufl.edu/discover-fish/sharks/shark-biology/.

———. "Shark Attack Numbers Remained 'Extremely Low' in 2020, but Fatalities Spiked." Posted January 25, 2021, accessed February 19, 2021. https://www.floridamuseum.ufl.edu/science/shark-attacks-extremely-low-in-2020-fatalities-spiked/

Foer, Joshua. "It's Time for a Conversation: Breaking the Communication Barrier between Dolphins and Humans." *National Geographic*, May 2015, 30–54.

Fogarty, Nicole D., and Kristen L. Marhaver. "Coral Spawning, Unsynchronized: Breakdown in Coral Spawning May Threaten Coral Reef Recovery." *Science 365*, no. 6457 (2019): 987–988.

Frank, Ron, William N. Eschmeyer, and Ray Van der Laan (eds.). *Catalog of Fishes: Genera, Species, References*. 2018. Updated January 11, 2021. http://researcharchive.calacademy.org/research/ichthyology/catalog/fishcatmain.asp.

Frost, Emily. "Improving Grades for the Mesoamerican Barrier Reef." *Smithsonian*. Last modified March 2018, accessed July 22, 2019. https://ocean.si.edu/ocean-life/plants-algae/improving-grades-mesoamerican-reef.

Fugu Fukuji. "Menu." Accessed January 30, 2019. https://www.fukuji.jp.

Garcia, Eden, and Karie Holterman. "Calabash Caye, Turneffe Islands Atoll, Belize." In *CARICOMP Caribbean Coral Reef, Seagrass and Mangrove Sites*, edited by Björn Kierfve, vol. 3, 67–78. Paris: UNESCO, 1998.

Gaskill, Melissa. "Diving the World Heritage List." *Alert Diver 33*, no. 4 (Fall 2017): 86–89.

———. "Watching Wildlife from Space." *Alert Diver 35*, no. 1 (Winter 2019): 18–20.

Georgia Aquarium. Accessed January 30, 2019. https://www.georgiaaquarium.org.

Ghiselin, Michael T. "Foreword." In Charles Darwin, *The Structure and Distribution of Coral Reefs.* Tucson: University of Arizona Press, 1984 (originally published in 1842).

Geophysical Fluid Dynamics Laboratory, National Oceanic and Atmospheric Administration. "Global Warming and Hurricanes." Last modified September 23, 2020, accessed November 2, 2020. http://gfdl.noaa.gov/global-warming-and-hurricanes/.

Godfrey-Smith, Peter. *Other Minds: The Octopus, the Sea, and the Deep Origins of Consciousness.* New York: Farrar, Strauss and Giroux, 2016.

González, Arturo H., Carmen Rojas Sandoval, Eugenio Acevez Núñez, Jerónimo Avilés Olguín, Santiago Analco Ramírez, Octavio del Río Lara, Pilar Luna Erreguerena, Adriana Velázquez Morlet, Wolfgang Stinnesbeck, Alejandro Terrazas Mata, and Martha Benavente San Vicente. "Evidence of Early Inhabitants in Submerged Caves in Yucatan, Mexico." *Underwater and Maritime Archeology in Latin America and the Caribbean,* edited by Margaret Leshikar-Denton and Pilar Luna Erreguerena, 127–142. Walnut Creek, CA: Left Coast Press, 2008.

Greenberg, Jerry, and Idaz Greenberg. *The Living Reef.* Miami: Seahawk Press, 1987.

Green Climate Fund. "Belize Country Strategic Framework Including a Country Programme for Engagement with the Green Climate Fund." Final Version, June 2019. Posted August 12, 2019, accessed February 8, 2020. https:/www.acclimatise.uk.com/wp-content/uploads/019/08/BELIZE-CP-Final-draft-version_12.08.19-CLEAN.pdf.

Greenfield, David A., and Teresa A. Greenfield. "Habitat and Resource Partitioning between Two Species of *Acanthern Hemoria* (Pisces: Cheenopsidae), with Comments on the Chaos Hypothesis." In Smithsonian Contributions to the Marine Sciences, no. 12, *The Atlantic Barrier Reef Ecosystem at Carrie Bow Cay, Belize, I: Structure and Communities,* edited by Klaus Rützler and Ian G. Macintyre, 497–507. Washington, DC: Smithsonian Institution Press, 1982.

Green Globe. "About." Accessed August 22, 2019. https://greenglobe.com/about/.

Greshko, Michael. "Ice Age Predators Found Alongside Oldest Human in Americas." *National Geographic.* Last modified August 25, 2017, accessed January 27, 2021. https://www.nationalgeographic.com/news/2017/08/ice-age-fossils-underwater-cave-bears-humans-science/.

Gress, Erika, Joshua D. Voss, Ryan J. Eckert, Gwilym Rowlands, and Dominic A. Andradi-Brown. "The Mesoamerica Reef." In *Mesophotic Coral Ecosystems,* edited by Yossi Loya, Kimberley A. Puglise, and Tom C. L. Bridge, 71–84. Cham, Switzerland: Springer Nature Switzerland AG, 2019.

The Guardian. "American Leaders Should Read Their Official Climate Science Report." US edition online. Last modified November 27, 2017. https://www.theguardian.

com/environment/climate-consensus-97-per-cent/2017/nov/27/american-leaders-should-read-their-official-climate-science-report.

———. "Nobel Prize in Physiology or Medicine 2012: As It Happened." US edition online. Last modified October 8, 2012. https://www.theguardian.com/science/blog/2012/oct/08/nobel-prize-2012-live-medicine-physiology.

———. "Tropical Storm Iota May Bring More Damage to Caribbean after Eta." Posted November 14, 2020, accessed November 14, 2020. http://theguardian.com/world/2020/nov/14/tropical-storm-iota-may-bring-more-damade-to-caribbean-after-eta.

———. "Warming of Oceans Due to Climate Change Is Unstoppable, Say US Scientists." US online edition. Last modified November 29, 2017. https://www.theguardian.com/environment/2015/jul/16/warming-of-oceans-due-to-climate-change-is-unstoppable-say-us-scientists#:~:text=Warming%20of%20oceans%20due%20to%20climate%20change%20is%20unstoppable%2C%20say%20US%20scientists,-This%20article%20is&text=The%20warming%20of%20the%20oceans,climate%20scientists%20said%20on%20Thursday.

Gutzler, Benjamin C., Mark J. Butler IV, and Donald C. Behringer. "Casitas: A Location-Dependent Ecological Trap for Juvenile Caribbean Spiny Lobsters, *Panulirus argus*." *ICES Journal of Marine Science* 17, no. suppl_1 (July 2015): i177–i184.

Guy, Michael, Hollie Silverman, and Judson Jones. "Hurricane Eta Rapidly Intensified and Is Now a Major Category 4 Hurricane Nearing Landfall." CNN. Posted and accessed November 2, 2020. http://cnn.com/2020/1/02/weather/tropical-storm-eta-Monday/index.html.

Hanlon, Roger. "Nature's Best and Fastest Camouflage." *Alert Diver* 33, no. 3 (Fall 2017): 44–47.

Hajovsky, Ric. "How Cozumel's Tourism Began." Last modified 2011, accessed March 14, 2019. http://everythingcozumel.com/cozumel-history/cozumels-tourism-industry-began/.

Hamanasi Adventure and Dive Resort. *Sustainable from the Ground Up* (Dangriga, Stann Creek, Belize: Hamanasi Adventure and Dive Resort, 2015).

Hardin, Garrett. "The Tragedy of the Commons." *Science, 162*, no. 3859 (December 13, 1968): 1243–1248.

Harte Research Institute for Gulf of Studies, Texas A&M University Corpus Christi. "Invasive Species." Edited by Fabio Moretzsohn, J. A. Sánchez Chávez, and John Wesley Turnell, Jr. GulfBase Resources Database for Gulf of Mexico Research, Harte Research Institute for Gulf Studies, 2004.

Healthy Reefs Initiative. "2016 Eco Audit." Accessed August 7, 2019. https://www.healthyreefs.org/cms/latest-reports/.

Helgason, Nicole. "Coral Bleaching Hits Mesoamerican Reef." Reef Divers. Last modified October 17, 2017, accessed July 22, 2019. https://reefdivers.io/coral-bleaching-hits-hard/4666.

Herrmann, Richard. "Tuna." *Alert Diver* 34, no. 1 (Winter 2018): 14–17.

Howard, Jenny. "Why Shark Attacks Are More Common in the Atlantic than the Pacific." *National Geographic.* Last modified July 2, 20919, accessed July 12, 2019. https://www.nationalgeographic.com/animals/2019/07/shark-attacks-atlantic-ocean/.

Iliffe, Thomas M. "Anchialine Caves and Cave Fauna of the World." *National Geographic.* Last modified March 2010, accessed August 30, 2017. http://magma.nationalgeographic.com/ngm/0310/feature4/index.htm

Index Mundi. "Guatemala Literacy." Accessed July 17, 2019. https://www.indexmundi.com/guatemala/literacy.html.

Island Expeditions. "Lighthouse Reef: Belize's First Marine Protected Area." Accessed August 11, 2019. https://www.islandexpeditions.com/belize-vacations-blog/lighthouse-reef-belizes-first-marine-protected-area.

Johnson, Paul G., and Barry A. Vittor. "Segmented Worms." In *Coral Reefs; A Guide to the Common Invertebrates and Fishes of Bermuda, the Bahamas, Southern Florida, the West Indies, and the Caribbean Coast of Central and South America,* edited by Eugene H. Kaplan, 134–140. Norwalk, CT: The Easton Press, 1982.

Johnsson, R., E. Neves, G.M.O. Franco, and F. L. Silveira. "The Association of Two Gall Crabs (*Brachyura: Cryptochiridae*) with the Reef-Building Coral *Siderastrea stellata* Verrill, 1868." *Hydrobiologia 559,* no. 1 (2006): 379–84.

Kairo, Moses. United Nations Environment Programme: The Caribbean Environment Programme. "Marine Invasive Species." Last modified 2003, accessed September 3, 2017. http://www.issg.org/database/species/reference_files/kairo.

Kaplan, Eugene H. *Coral Reefs: A Guide to the Common Invertebrates and Fishes of Bermuda, the Bahamas, Southern Florida, the West Indies, and the Coast of Central and South America.* Norwalk, CT: The Easton Press, 1982.

Kaufman, Les, and John Tschirky, "Living with the Sea: Local Efforts Buffer Effects of Global Change." Arlington, VA: Conservation International, Science and Knowledge Division, 2010.

Kiefvre, Björn (ed.). *CARICOMP Caribbean Coral Reef, Seagrass and Mangrove Sites.* vol. 3. Paris: UNESCO, 1998.

Kilfeather, Siobhan Marie. *Dublin: A Cultural History.* Oxford: Oxford University Press, 2005.

Kinzie, Robert III. "Soft Corals." In *Coral Reefs: A Guide to the Common Invertebrates and Fishes of Bermuda, the Bahamas, Southern Florida, the West Indies, and the Caribbean Coast of Central and South America,* edited by Eugene H. Kaplan, 86–95. Norwalk, CT: The Easton Press, 1982.

Knoema. "Quintana Roo: Total Population." Posted January 2021, accessed February 14, 2021. http://knoema.com/atlas/Mexico/Quintana-Roo/topics/Demographics/Key-Indicators/Population.

Knowlton, Nancy. *Citizens of the Sea: Wondrous Creatures from the Census of Marine Life.* Washington, DC: National Geographic, 2010.

———. "Ocean Optimism." *Alert Diver 34,* no. 1 (Winter 2018): 82–85.

———. "Sea Urchin Recovery from Mass Mortality: New Hope for Caribbean Coral Reefs?" in *Proceedings of the National Academy of Sciences of the United States,* April 24, 2001.

Kolbert, Elizabeth. "Going Negative." *New Yorker,* November 20, 2017, 64–73.
——. "Now You See It." *New Yorker,* October 15, 2018, 97–99.
——. "Unnatural Selection: What Will It Take to Save the World's Reefs and Forests?" *New Yorker,* April 18, 2016, 22–29.
Kokemuller, Jill. "What Are Shrimps' Prey?" Last modified August 6, 2018, accessed July 4, 2019. https://sciencing.com/shrimps-prey-8309570.html.
Kormann, Carolyn. "The Widening Gyre: A Young Entrepreneur Takes on Ocean Pollution." *New Yorker,* February 4, 2019, 42–49.
Kuczaj, Stan A., and Holli C. Eskelinen. "Why Do Dolphins Play?" *Animal Behavior and Cognition 1,* no. 2 (2014): 113–127.
Kuczaj, Stan A. II, and Deidre Yeater. "Dolphin Imitation: Who, What, When, and Why?" *Aquatic Mammals 32,* no. 4 (2006): 413–422.
Kwon, Diana. "What Makes Our Brains Special?" *Scientific American.* Last modified November 24, 2015, accessed July 11, 2019. https://www.scientificamerican.com/article/what-makes-our-brains-special/.
Lai, Bun. "How (and Why) to Eat Invasive Species," also published as "Invasive Species Menu of a World-Famous Chef," *Scientific American.* Last modified September 1, 2013, accessed September 2, 2017. http://www.scientificamerican.com/article/invasive-species-menu-of-a-world-famous-chef.
Lamar University. "Marine Critters." Accessed February 1, 2019. https://www.lamar.edu/marine-critters.
Larson, Ronald J. "Medusae (Cnidaria) from Carrie Bow Cay, Belize." in Smithsonian Contributions to the Marine Sciences, no. 12, *The Atlantic Barrier Reef Ecosystem at Carrie Bow Cay, Belize, I: Structure and Communities,* edited by Klaus Rützler and Ian G. Macintyre, 252–258. Washington, DC: Smithsonian Institution Press, 1982.
Laughing Bird. "Welcome to Laughing Bird Caye National Park in Belize." Last modified 2019, accessed August 11, 2019. https://www.laughingbird.org/park1.html.
Leatherman, Courtney. "Interview: Stephanie Wear." *Nature Conservancy,* March–April 2013, 26–27.
Leshikar-Denton, Margaret, and Pilar Luna Erreguerena. "The Foundations of Underwater and Maritime Archeology in Latin America and the Caribbean." In *Underwater and Maritime Archeology in Latin America and the Caribbean,* edited by Margaret Leshikar-Denton and Pilar Luna Erreguerena, 25–53. Walnut Creek, CA: Left Coast Press, 2008.
Lessios, Harilaos. "The Great *Diadema antillarium* Die-Off: 30 Years Later." *Annual Review of Marine Science 2016 8* (June 5, 2015): 267–283.
Lessios, Harilaos, D. Ross Robertson, and John D. Cubit. "Spread of *Diadema* Mass Mortality through the Caribbean." *Science 226* (November 1984): 335–337.
Lewis, Robert. "Manta Ray." *Encyclopaedia Britannica.* Accessed July 14, 2019. https://www.britannica.com/animal/manta-ray/.
Ley-Cooper, Kim. *Sustainability of Lobster* Panulirus argus *Fisheries in Marine Protected Areas in South-eastern Mexico.* Doctoral thesis. Curtin University, Perth, Australia, submitted December 2015.

Lopez, Oscar. "'Coral Bleaching Is Getting Worse . . . but the Biggest Problem Is Pollution.'" *The Guardian*. Last modified December 27, 2017, accessed August 13, 2019. https://www.theguardian.com/environment/2017/dec/27/coral-bleaching-is-getting-worse-but-the-biggest-problem-is-pollution.

Loya, Yossi, Kimberley A. Puglise, and Tom C. L. Bridge (eds.). *Mesophotic Coral Ecosystems*. Cham, Switzerland: Springer Nature Switzerland AG, 2019.

Luna Erreguerena, Pilar. "The Submerged Cultural Heritage in Mexico." In *Underwater and Maritime Archeology in Latin America and the Caribbean*, edited by Margaret Leshikar-Denton and Pilar Luna Erreguerena, 55–65. Walnut Creek, CA: Left Coast Press, 2008.

Luther, Carol. "Manatee Tours in Belize." *USA Today*. Accessed March 12, 2019. https://manatee-tours-in-belize-11347.html.

MacGregor, Jeff. "Taming the Lionfish." *Smithsonian*, June 2018: 24–41.

Macintyre, Ian G., Klaus Rützler, James N. Norris, and Kristian Fauchald. "A Submarine Cave near Columbus Cay, Belize: A Bizarre Cryptic Habitat." In Smithsonian Contributions to the Marine Sciences, no. 12, *The Atlantic Barrier Reef Ecosystem at Carrie Bow Cay, Belize, I: Structure and Communities*, edited by Klaus Rützler and Ian G. Macintyre, 126–141. Washington, DC: Smithsonian Institution Press, 1982.

Macrotrends. "Guatemala." Posted January 2021, accessed February 12, 2021. https://www.macrotrends.net/countries/GTM/gutemala/population-growth/rate.

———. "Honduras." Posted January 2021, accessed February 12, 2021. https://www.macrotrends.net/countries/HON/honduras/population-growth-rate/.

———. "Mexico." Posted January 2021, accessed February 17, 2021. https://www..macrotrends.net/countries/MEX/mexico/population.

Main, Douglas. "Why Koko the Gorilla Mattered." *National Geographic*. Last modified June 21, 2018, accessed July 8, 2019. https://www.news.nationalgographic.com/2018/06/gorillas-koko-sign-language-culture-animals.

Marine Education Society of Australasia. "Crustaceans." Last modified June 25, 2019, accessed June 25, 2019. https://www.mesa.edu.au/crustaceans/.

Martin, Judith. "A Ring around the Mood Market." *Washington Post*, November 24, 1975, B9.

Marturana, Amy. "This Is What Actually Happens to Your Skin When You Get a Tan." *Self*. Last modified June 8, 2016, accessed August 20, 2019. https://www.self.com/story/this-is-what-actually-happens-to-your-body-when-you-get-a-tan.

Mateo, Miguel A., Just Cebrián, Kenneth Dunton, and Troy Mutchler. "Carbon Flux in Seagrass Ecosystems." Chapter Seven in *Seagrasses: Biology, Ecology and Conservation*, edited by A.W. D. Larkum, R. J. Orth, and C. M. Duarte, 159–192. The Netherlands: Springer, 2006.

McField, Melanie. *Mesoamerican Reef: An Evaluation of Ecosystem Health, Healthy Reefs for Healthy People, 2015 Report Card*. (Playa del Carmen, Mexico: Healthy Reefs for Healthy People, 2015). Accessed March 13, 2019. https://www/healthyreefs.org/cms/wp-content/uploads/2015/05/MAR-EN-small.pdf.

———. *Report Card for the Mesoamerican Reef: An Evaluation of Ecosystem Health 2010*.

(Playa del Carmen, Mexico: Healthy Reefs for Healthy People, 2010). https://www/healthyreefs.org/cms/wp-content/uploads/.

———. *Report Card for the Mesoamerican Reef 2016.* (Playa del Carmen, Mexico: Healthy Reefs for Healthy People, 2016).

McField, Melanie, Patricia Kramer, Ana Giró Petersen, Mélina Soto, Ian Drysdale, Nicole Craig, and Marisol Rueda Flores, *Mesoamerican Reef Report Card 2020* (Playa del Carmen, Mexico: Healthy Reefs for Healthy People). Posted February 2020, accessed June 16, 2020. https://www.healthyreefs.org/cms/wp-content/uploads/2020/02/2020-Report-Card-MAR.pdf.

McLenachan, Loren. "Documenting Loss of Large Trophy Fish from the Florida Keys with Historical Photographs." *Conservation Biology 23,* no. 3 (2009): 636–643.

Meadows, Robin. "Overfishing Implicated in Sea Urchin Epidemics." *Conservation Magazine.* Last modified July 29, 2008, accessed February 11, 2019. https://www.conservationmagazine.org/2008/07/overfishing-implicated-in-sea-urchin-epidenics.

Mervosh, Sarah. "Carnival Cruises to Pay $20 Million in Pollution and Cover-Up Case." *New York Times.* Posted June 4, 2019, accessed October 30, 2020. http://nytimes.com/2019/06/04/businesss/carnival-crruise-pollution.html.

Mexico News Daily. "New Protected Area to Be Mexico's Biggest." Last modified December 3, 2016, accessed August 12, 2019. https://mexiconewsdaily.com/news/new-protected-area-to-be-mexicos-biggest/.

Milman, Oliver. "Below Bermuda, the Quest to Map the Damage We Are Doing to the Deep Sea." *The Guardian.* Last modified August 19, 2016. https://www.theguardian.com/environment/2016/aug/17/ocean-research-marine-life-bermuda-coral-reefs-nekton-triton-vessel.

Montgomery, Sy. *The Soul of an Octopus: A Surprising Exploration into the Wonder of Consciousness.* New York: Atria Books, 2015.

Muzik, Katherine. "*Octocorallis* (Cnidaria) from Carrie Bow Cay, Belize." In Smithsonian Contributions to the Marine Sciences, no. 12, *The Atlantic Barrier Reef Ecosystem at Carrie Bow Cay, Belize, I: Structure and Communities,* edited by Klaus Rützler and Ian G. Macintyre, 303–310. Washington, DC: Smithsonian Institution Press, 1982.

National Geographic. "Hammerhead Sharks." Accessed July 17, 2019. https://www.nationalgeographic.com/animals/fish/group/hammerhead-sharks/.

———. "Nudibranchs." Accessed July 3, 2019. https://www.nationalgeographic.com/animals/invertebrates/group/nudibranchs/.

———. "Sea Turtles Match Breathing to Diving Depths?" Last modified November 23, 2010, accessed February 19, 2019. http://www.video.nationalgeographic.com/video/news/00000144-0a34-d3c b.a9bc-763df1360000.

———. "Tiger Shark." Accessed July 16, 2019. https://www.nationalgeographic.com/animals/fish/t/tiger-shark/.

———. "Whale Shark." Accessed July 18, 2019. https://www.nationalgeographic.com/animals/fish/w/whale-shark/.

National Oceanic and Atmospheric Administration [website]. Last modified January 15, 2016, accessed December 5, 2016. https://www.noaa.gov.

———. "Black Grouper." Accessed January 31, 2019. https://www.fisheries.noaa.gov/species/black-grouper.

———. "Caribbean Spiny Lobster." Accessed June 27, 2019. https://www.fisheries.gov/species/caribbean-spiny-lobster.

———. "How Does Sand Form?" Last modified April 9, 2020, accessed May 30, 2020. https://www.oceanservice.noaa.gov/facts/sand.html.

———. "How Many Species Live in the Ocean?" Last modified January 7, 2020, accessed January 29, 2020. https://www.oceanservice.noaa.gov/facts/ocean-species.html.

——— "Invasive Cup Coral." Flower Gardens Banks, National Marine Sanctuary. Last modified July 31, 2017, accessed September 3, 2017. https://flowergardens.noaa.gov/invasivecupcoral.html.

———. "In What Types of Water Do Corals Live?" Last modified January 7, 2020, accessed May 30, 2020. https://www.oceanservices.noaa.gov/facts/coralwaters.html.

———. "Nassau Grouper." Accessed January 31, 2019. https://www.fisheries.noaa.gov/species/nassau-grouper.

———. "Sea Turtles." Accessed February 18, 2019. https://www.fisheries.noaa.gov/sea-turtles.

———. "What Is a Gyre?" Accessed August 16, 2019. https://oceanservice.noaa.gov/facts/gyre.html.

———. "What's the Difference between Dolphins and Porpoises?" Accessed July 7, 2019. https://oceanservice/noaa.gov/facts/dolphin_porpoise.html.

National Wildlife Federation. "Sea Turtle: Hawksbill." Posted 2021, accessed February 14, 2021. http://wwf.orrg/Educational-Resources/Wildlife-Guide/Reptiles/Sea-Turtles/Hawksbill-Sea-Turtle.

Nelson, Bryan. "12 Animals with the Longest Gestation Period." Mother Nature Network. Last modified April 28, 2010, accessed July 13, 2019. https://www.mnn.com/earth-matters/animals/photos/12-animals-with-the-longest-gestation-period/sharks.

New York Times. Ruth Gates obituary. Last modified November 5, 2018, accessed August 14, 2019. https://www.nytimes.com/2018/11/05/obituaries/ruth-gates-dead-marine-biologist-who-championed-coral.html.

Norris, James N., and Katina E. Bucher. "Marine Algae and Seagrass from Carrie Bow Cay, Belize." In Smithsonian Contributions to the Marine Sciences, no. 12, *The Atlantic Barrier Reef Ecosystem at Carrie Bow Cay, Belize, I: Structure and Communities,* edited by Klaus Rützler and Ian G. Macintyre, 167–223. Washington, DC: Smithsonian Institution Press, 1982.

Norris, James N., and William Fenical. "Chemical Defense in Tropical Marine Algae." In Smithsonian Contributions to the Marine Sciences, no. 12, *The Atlantic Barrier Reef Ecosystem at Carrie Bow Cay, Belize, I: Structure and Communities,* edited by

Klaus Rützler and Ian G. Macintyre, 417–431. Washington, DC: Smithsonian Institution Press, 1982.

Nuñez, David. "Mexico—Quintana Roo—The Vigia Chico Fishing Cooperative." The EcoTipping Points Project: Models for Success in a Time of Crisis. Last modified November 2006, accessed July 31, 2019. http://www.ecotippingpoints.org/our-stories/indepth/mexico-quintana-roo-vigia-chico.html.

Ocean Cleanup. "Interceptor 011 in Place after Extreme Flooding in Jakarta." Posted January 22, 2020, accessed February 23, 2021. https://theoceancleanup/general/interceptor-011-in-place-after-extreme-flooding-in-jakarta.

———. "The Ocean Cleanup Successfully Catches Plastic in the Great Pacific Garbage Patch." Posted October 2, 2019, accessed February 23, 2021. https://theoceancleanup.com/updates/the-ocean-cleanup-successfully-catches-plastic-in-the-great-pacific-garbage-patch.

———. "Thailand Interceptor Project Kicks off with Signing of MOU." Posted February 3, 2021, accessed February 23, 2021. https://theoceancleanup.com/updates/thailand-interceptor-project-kicks-off-with-signing-of-mou.

———. "Turning Trash into Treasure: The Ocean Cleanup Sunglasses." Posted October 24, 2020, accessed February 23, 2021. https://theoceancleanup.com/general/turning-trash-into-treasure-the-ocean-cleanup-sunglasses.

Ogden, John C. "Book Review: *Seagrasses, Ecology and Conservation.*" *Marine Ecology* 27, no. 4, edited by A. W. D. Larkum, R. J. Orth, and C. M. Duarte, 431–432. The Netherlands: Springer, 2006.

Olson, Ted. "Invasion of the Jellyfish." *Scholastic Action.* Last modified February 2018, accessed January 13, 2019. https://action.scholastic.com/issues/2017-18/020118/invasion-of-the-jellyfish—jellyfish-for-dinner.html#720L.

Orbach, Michael, and Leah Karrer. "Marine Managed Areas: What, Why, and Where." Arlington, VA: Science and Knowledge Division, Conservation International, 2010.

Osborn, Liz. "Number of Species Identified on Earth." Last modified 2019, accessed January 28, 2019. https://www.currentresults.com/Environment-Facts/Plants-Animals/number-species.php.

Peake, Jonathan, Alex K. Bogdanoff, Craig A. Layman, Bernard Castillo, Kynoch Reale-Munroe, Jennifer Chapman, Kristen Dahl, William F. Patterson III, Corey Eddy, Robert D. Ellis, Meaghan Faletti, Nicholas Higgs, Michelle A. Johnston, Roldan C. Muñoz, Vera Sandel, Juan Carlos Villasenor-Derbez, and James A. Morris Jr. "Feeding Ecology of Invasive Lionfish (*Pterois volitans* and *Pterois miles*) in the Temperate and Tropical Western Atlantic." *Biological Invasions.* Posted April 11, 2018, accessed January 21, 2019. https://doi.org/10.1007/s10530-018-1720-5.

Pelikan, Kellie C. "The Effects of Petroleum Pollutants on Sea Urchins Reproduction and Development." Master's thesis. Nova Southeastern University. Submitted December 7, 2015, accessed February 9, 2019. https://nsuworks.nova.edu/occ_stu-etd/401.

Petsko, Emily. "Tackling a Triple Threat: Belize Banned Bottom Trawling, Offshore Drilling, and Now Gillnets." Oceana, December 8, 2020. http://oceana.org/blog/

tackling-a-triple-threat-belize-banned-bottom-trawling-offshore-drilling-and-now-gillnets.

Polyak, Victor J., Bogdan P. Onac, Joan J. Fornós, Carling Hay, Yemane Asmerom, Jeffrey A. Dorale, Joaquín Ginés, Paola Tuccimei, and Angel Ginés. "A Highly Resolved Record of Relative Sea Level in the Western Mediterranean Sea during the Last Interglacial Period." *Nature Geoscience,* 11 (2018): 11, 860–864. https://doi.org/10.1038/s41561-018-0222-5.

Pro México. "Mexico Seafood Industry." Accessed January 22, 2020. http://www.gob.mx/cms/uploads/attachment/file/637431/seafood-industry.pdf.

Pro México Trade and Investment. "Quintana Roo." Accessed March 12, 2019. http://mim.promexico.gob.mx/work/models/mim/ . . . /PDF/FE-QUINTANA_ROO.

Reefguide.org. "Florent's Guide to the Florida, Bahamas & Caribbean Reefs." Accessed February 8, 2019. https://reefguide.org/carib/index22.html.

Rice, Mary W., and Ian G. Macintyre. "Distribution of Sipuncula in the Coral Reef Community, Carrie Bow Cay, Belize." In Smithsonian Contributions to the Marine Sciences, no. 12, *The Atlantic Barrier Reef Ecosystem at Carrie Bow Cay, Belize, I: Structure and Communities,* edited by Klaus Rützler and Ian G. Macintyre, 311–320. Washington, DC: Smithsonian Institution Press, 1982.

Rojas Sandoval, Carmen, Arturo H. González, Alejandro Terrazas Mata, and Martha Benavente Sanvicente. "Mayan Mortuary Deposits in the Cenotes of Yucatán and Quintana Roo, Mexico." In *Underwater and Maritime Archeology in Latin America and the Caribbean,* edited by Margaret Leshikar-Denton and Pilar Luna Erreguerena, 143–153. Walnut Creek, CA: Left Coast Press, 2008.

Rolex.org. "Arturo González: In Search of the First Americans." Posted 2008, accessed January 18, 2020. http://rolex.org/eng/rolex-awards/exploration/arturo-gonzalez.

Roser, Max. "Child and Infant Mortality." Our World in Data. Accessed July 9, 2019. https://ourworldindata.org/child-mortality.

Rützler, Klaus, and Ian G. Macintyre. "The Habitat Distribution and Community Structure of the Barrier Reef Complex at Carrie Bow Cay, Belize." In Smithsonian Contributions to the Marine Sciences, no. 12, *The Atlantic Barrier Reef Ecosystem at Carrie Bow Cay, Belize, I: Structure and Communities,* edited by Klaus Rützler and Ian G. Macintyre, 21–45. Washington, DC: Smithsonian Institution Press, 1982.

———. "Preface." In Smithsonian Contributions to the Marine Sciences, no. 12, *The Atlantic Barrier Reef Ecosystem at Carrie Bow Cay, Belize, I: Structure and Communities,* edited by Klaus Rützler and Ian G. Macintyre, i–xiv. Washington, DC: Smithsonian Institution Press, 1982.

Saint Louis Zoo. "Things to See and Do: Stingrays at Caribbean Cove." Accessed July 14, 2019. https://www.stlzoo.org/visit/thingstoseeanddo/stingraysatcaribbeancove/southernstingrayfacts.

Samonte, Giselle, Leah Bunce Karrer, and Michael Orbach. *People and Oceans: Managing Marine Areas for Human Well-Being.* Arlington, VA: Science and Knowledge Division, Conservation International, 2010.

San Diego Zoo. "Dolphin." Accessed July 8, 2019. https://animals.sandiegozoo.org/animals/dolphin.

San Pedro Sun. "Belize Advances in Climate Change Adaptation Planning and Disaster Risk Management with Support from UB and IDB." Posted November 9, 2019, accessed February 8, 2020. https://www.sanpedrosun.com/government/2019/11/09/belize-advances-in-climate-change-adaptation-planning-and-disaster-risk-management-with-support-from-ub-and-idb.

Schwartz, Matthew S. "Key West to Ban Popular Sunscreen Ingredients to Protect Coral Reef." National Public Radio. Last modified February 6, 2019, accessed August 20, 2019. https://www.npr.org/2019/02/06/691913378/key-west-votes-to-ban-popular-sunscreen-ingredients-to-protect-coral-reef.

Science Daily. "Scientists Find Stable Sea Levels during Last Interglacial." Posted September 2018, accessed February 14, 2021. https://sciencedaily.com/releases/2018/09/180910111314.htm.

Sea Turtle Conservancy. "Information about Sea Turtles: Frequently Asked Questions." Accessed February 19, 2019. https://conserveturtles.org/information-sea-turtles.

Sea World Parks and Entertainment. "All about Bottlenose Dolphins." Accessed July 9, 2019. https://seaworld.org/animals/all-about/bottlenose-dolphins/longevity/.

———. "All about Sea Turtles." Accessed February 18, 2019. https://seaworld.org/animals/all-about/sea-turtles.

Share the Beach. "Alabama Sea Turtle Nesting Facts." Accessed February 25, 2019. https://www.alabamaseaturtles.com/nesting-sea-turtle-statistics.

Shark Research Institute. "Catching Some Rays in Mexico." Accessed July 14, 2019. https://www.sharks.org/shark-research-institute-blogs/blogs/science-blog/catching-rays-mexico.

Shinn, Eugene A., R. B. Halley, J. H. Hudson, B. Lidz, D. M. Robbin, and I. G. Macintyre. "Geology and Sediment Accumulation Rates at Carrie Bow Cay, Belize." In Smithsonian Contributions to the Marine Sciences, no. 12, *The Atlantic Barrier Reef Ecosystem at Carrie Bow Cay, Belize, I: Structure and Communities,* edited by Klaus Rützler and Ian G. Macintyre, 63–75. Washington, DC: Smithsonian Institution Press, 1982.

Siegel, Jules. *Cancún User's Guide.* Quintana Roo, Mexico: The Communication Company, 2006, 204.

Simon, Matt. "Absurd Creature of the Week: This Fish Swims up a Sea Cucumber's Butt and Eats Its Gonads," *Wired,* February 21, 2014. Accessed March 26, 2019. https://www.wired.com/2014/02/absurd-creature-of-the-week-pearlfish/.

Slat, Boyan. "Into the Twilight Zone." The Ocean Cleanup. Last modified August 16, 2019, accessed August 16, 2019. https://theoceancleanup.com/updates/.

Smith, Adam. "The Science and Art of Reef Restoration." World Economic Forum. Last modified July 17, 2018, accessed February 8, 2019. https://www.weforum.org/agenda/2018/07/the-science-and-art-of-reef-restoration.

Smith, Julian. "Bracing for Impact." *Nature Conservancy,* Winter 2018, 50–59.

Southern Environmental Association. "Research." Posted 2020, accessed February 7, 2020. https://www.seabelize.org/research/.

Sponge Docks. "Tarpon Springs Culture and History." Accessed January 21, 2019. https://spongedocks.net/tarpon-springs-history.htm.

Sportalsub.net. "Danish Stig Severinsen Sets New Guinness World Record for Dynamic Apnea at Sea in Mexico." Posted December 24, 2020, accessed February 16, 2021. http://www.sportalsub.net/stig-severinsen-record-guiness-2020/.

Swain, Frank. "World's Fastest Shark Gets a Burst of Speed from Shape-Shifting Skin." *New Scientist*. Last modified March 4, 2019, accessed July 4, 2019. https://www.newscientist.com/article/2195435-worlds-fastest-shark-gets-a-burst-of-speed-from-shape-shifting-skin/.

Taylor, Gordon T., and Richard D. Bray. "Brittle or Serpent Stars." In *Coral Reefs; A Guide to the Common Invertebrates and Fishes of Bermuda, the Bahamas, Southern Florida, the West Indies, and the Caribbean Coast of Central and South America*, edited by Eugene H. Kaplan, 177–186. Norwalk, CT: The Easton Press, 1982.

Telford, Malcolm. "Shrimps, Lobsters, and Crabs." In *Coral Reefs: A Guide to the Common Invertebrates and Fishes of Bermuda, the Bahamas, Southern Florida, the West Indies, and the Caribbean Coast of Central and South America*, edited by Eugene H. Kaplan, 150–168. Norwalk, CT: The Easton Press, 1982.

Tennenhouse, Erica. "These Fishermen Helping Dolphins Have Their Own Culture." *National Geographic*. Last modified April 9, 2019, accessed July 8, 2019. https://www.nationalgeographic.com/animals/2019/04/dolphins-fishermen-brazil-culture/.

Than, Ker. "Jacques Cousteau Centennial: What He Did, Why It Matters." *National Geographic*. Last modified June 11, 2010, accessed September 1, 2017. https://www.news.nationalgeographic.com.

Thole, Rutger. "Five Shark Species Your Will Encounter Scuba Diving in the Caribbean Sea." Accessed July 13, 2019. https://rushkult.com/eng/scubamagazine/shark-species-found-in-the-caribbean-sea/.

Thompson, Edward H. "Atlantis Not a Myth." *Alpena Weekly Argus*, February 4, 1880.

Tidwell, James H., and Geoff L. Allan. "Fish as Food: Aquaculture's Contribution." *EMBO Reports 2*, no. 11, November 15, 2001, 958.

Tikkanen, Amy. "Bay Islands." *Encyclopaedia Britannica*. Accessed March 14, 2019. https://www.britannica.com/place/Bay-Islands.

Toonen, R. "Aquarium Invertebrates: Housing an Octopus." *Advanced Aquarist*. Accessed July 2013. https://reefs.com/magazine/aquarium-invertebrates-housing-an-octopus/.

Truman, Ryan. "Caves of the Yucatán." Webpage created for Earlham College Physical Geology 2004. Last modified April 20, 2004, accessed August 29, 2017. http://legacy.earlham.edu/~truman/yucatancave.html.

University of California at Santa Barbara. "UCSB Science Line." Last modified June 6, 2001, accessed January 24, 2019. https://www.scienceline.ucsb.edu.

U.S. Geological Survey. "Taming the Lion(fish)." Posted July 27, 2016, accessed October 26, 2020. http://usgs.gov/news/taming-lionfish.

U.S. Whales. "How Do Dolphins Give Birth?" Accessed July 7, 2019. https://us.whales. org/whales-dolphins/how-do-dolphins-give-birth/.

———. "How Do Dolphins Sleep?" Accessed July 8, 2019. https://us.whales.org/ whales-dolphins/how-do-dolphins-sleep/.

Van Giersen, Lena, Peter B. Kilian, Corey A. H. Allard, and Nicolas W. Bellono. "Molecular Basis of Chemotactile Sensation in Octopus." *Cell 183*, no. 3 (October 29, 2020): 594–604.

Vásquez Yeomans, Lourdes, and Martha Elena Valdez Moreno. "Códigos de Barras de la Vida en Huevos y Larvas de Peces Costeros y Oceánicos de la Parte Norte del Sistema Arrecifal Mesoamericano (Caribe Mexicano)." 2009/12/15–2012/10/11. https://www.ecosur.mx/ecoconsulta/busqueda/detalles.php?id=43063&bdi=0.

Voss, Gilbert L., and Clyde F. E. Roper. "Cephalopod Class of Mollusks." *Encyclopaedia Britannica* online. Last modified March 3, 2017, accessed September 1, 2017. https:// www.britannica.com/animal/cephalopod.

Washington Post. John C. Lilly obituary. "Inventor Studied Dolphin Communications." Last modified October 4, 2001, accessed July 10, 2019. https://www.washington-post.com/archive/local/2001/10/04/john-c-lilly/594dc878–43ac-.

Wassilieff, Maggy, and Steve O'Shea. "Octopus and Squid—Feeding and Predation." Te Ara—the Encyclopedia of New Zealand. Last modified June 12, 2006, accessed September 1, 2017 (published June 12, 2006). http://www.TeAra.govt.nz/en/octo-pus-and-squid/page-3.

Waymer, Jim. "Lionfish vs. Moray Eel: No Contest." *Florida Today.* Last modified February 24, 2017, accessed September 4, 2017. https://www.floridatoday.com/story/ news/local/environment/2017.

Webster, Bayard. "Sea Urchin Deaths Puzzle Scientists." *New York Times.* Last modified October 12, 1984, accessed February 9, 2019. https://www.nytimes. com/1984/10/12/us/sea-urchin-deaths-puzzle-scientists.html.

Wei-Haas, Maya. "One of the Biggest Great White Sharks Feasting on a Sperm Whale in Rare Video." *National Geographic.* Last modified July 19, 2019, accessed July 26, 2019. https://www.nationalgeorgrphic.com/aanimals/2019/07/rare-footage-three-female-great-white-sharks.

Wildlife Conservation Society. "Government of Belize Expands Marine Protected Areas in Biodiverse Offshore Waters." Posted April 3, 2019, accessed February 15, 2021. http:// newsroom.wcs.org/News-Releases/articleType/ArticleView/ed/12150/ government-of-belize-expands-marine-protected-areas-in-biodiverse-offshore-waters.

Worldometers. "Belize Population." Accessed August 7, 2019. https://www.worldom-eters.info/world-population/belize-population/.

Worldometers. "Honduras Population." Accessed March 14, 2019. https://www.worl-dometters.info/world.population/honduras.population/.

World Population Review. "Belize Population." Accessed February 24, 2021. https://www.worldpopulationreview.com/countries/belize-population.

World Wildlife Fund. "Mesoamerican Barrier Reef." Last modified April 11, 2012, accessed March 9, 2019. https://www.worldwildlife.org/places/mesoamerican-reef.

———. "Our Work: Biodiversity." Posted 2019, accessed February 1, 2020. https://www.wwf.panda.org/our-work/biodiversity/biodiversity.

———. "Sea Turtle: Hawksbill." Posted 2021, accessed February 14, 2021. http://wwf.org/Educational-Resources/Wildlife-Guide/Reptiles/Sea-Turtles/Hawksbill-Sea-Turtle.

Yeginsu, Ceylan. "Where Cruise Ships Are Sent to Die." *New York Times,* October 30, 2020. Accessed October 30, 2020. http://nytimes.com/2020/10/30/travel/cruise-ships-scrapped.html.

Zachos, Elaina. "Why Are We Afraid of Sharks? There's a Scientific Explanation." *National Geographic.* Last modified June 27, 2019, accessed July 12, 2019. https://news.nationalgeographic.com/2018/01/sharks-attack-fear-science-psychology-spd/.

Zeilinski, Sarah. "What Preys on Humans?" *Smithsonian Magazine.* Last modified July 22, 2011, accessed July 16, 2019. https://www.smithsonianmag.com/science-nature/what-preys-on-humans-34332952/.

Zeldovich, Lina. "The Great Dolphin Dilemma." *Hakai Magazine.* Last modified February 5, 2019, accessed July 7, 2019. https://www.hakaimagazine.com/features/the-great-dolphin-dilemma.

INDEX

Although Sandy Sheehy was born in New York, after graduating from Vassar, she moved to Austin then Houston, Galveston, and Albuquerque. She now divides her time between Texas and New Mexico. Sheehy and her husband, historian and University of New Mexico professor emeritus Charles McClelland, spend several months a year traveling internationally.

Sheehy has written frequently on human relationships and the natural environment. Her work has appeared in *Town & Country, Forbes, House Beautiful, Self, Working Woman,* and *Money,* among other national magazines. She is the author of *Texas Big Rich,* a group portrait of the state's financially fortunate, and *Connecting: The Enduring Power of Female Friendship.* Her first novel, *Deserts of the Heart,* is a historical romance set in 1798 near Spanish Colonial Santa Fe. The present book, *Imperiled Reef: The Fascinating, Fragile Life of a Caribbean Wonder,* grew out of her love of scuba diving. "Drifting weightless above a coral head is the closest those of us alive today will ever come to visiting another planet," she explains.